G. N. Sobirova
S. S. Rakhimkhodjaev

Structure, design and technology of openwork fabrics production

G. N. Sobirova
S. S. Rakhimkhodjaev

Structure, design and technology of openwork fabrics production

ScienciaScripts

Cover image: www.ingimage.com

This book is a translation from the original published under ISBN 978-620-7-48429-4.

Publisher:
Sciencia Scripts
is a trademark of
Dodo Books Indian Ocean Ltd. and OmniScriptum S.R.L publishing group

120 High Road, East Finchley, London, N2 9ED, United Kingdom
Str. Armeneasca 28/1, office 1, Chisinau MD-2012, Republic of Moldova, Europe
Managing Directors: Ieva Konstantinova, Victoria Ursu
info@omniscriptum.com

Printed at: see last page
ISBN: 978-620-8-37007-7

Contents

Annotation

The work is devoted to the structure, design and production technology of openwork fabrics for the upper of footwear insoles. The existing ligating means are analysed. The leno means of reducing the tension of stockinette threads at the moment of transition of the eyes of leno galleries relative to stockinette threads is developed. The main parameters determining the structure of fabrics are considered, the regularities of fabric density changes in warp and weft, the height of warp and weft bending waves are obtained. Limiting maximum densities of warp and weft for plain weave and cambric weave are determined. The influence of weave rapport, number of transitions, linear density and type of weft yarns, warp and weft yarns tension on yarns processing in fabrics is investigated when constructing designs of falsely openwork weaves when producing fabrics. The technical calculation of fabrics is made and the technological process of preparation and production of falsely openwork fabrics is chosen. Physico-mechanical investigations of the produced samples of falsely openwork weave fabrics are carried out. When producing falsely openwork fabrics for shoe tops, the minimum breakage of the main threads, which are recommended to be set for the following optimal parameters: the tension of the main threads - 20 cN (per 1 thread); the size of the scalp - 15 mm.; the position of the scalp relative to the breast - (+25) mm. An algorithm and a programme for calculation of shoe fabrics according to the given thickness for insoles are developed. Physical-mechanical and hygienic properties of fabrics for shoe insoles are investigated.

Keywords: yarn, warp, weft, fabric, processing parameters, openwork, falsely openwork pattern.

GENERAL CHARACTERISATION OF WORK

One of the most important tasks of modern production of industrial enterprises to increase their efficiency on the basis of improving the quality indicator, which determines the economic and social development. When considering the issues of economic independence of the Republic, preparation for transition to market relations, one of the urgent problems is intensification of production, search for internal resources, full processing of raw materials into finished products, the possibility of entering the foreign market. In this regard, in the future the task is set not only to satisfy the need at the expense of local production, but also to increase the supply of finished products from Uzbekistan to other CIS countries, as well as to foreign countries.

Aim and objectives of the research. The aim of the work is to study the structure, design and production technology of openwork fabrics. In order to solve the set task it is planned:

- analytically investigate the structure of falsely openwork weaves;
- carrying out technological studies of falsely openwork fabrics;
- Study of physical and mechanical properties of falsely openwork fabrics;

Research methodology. To solve the set tasks in the work a complex method of research was used. In the theoretical part, geometric methods were used to study the structure of falsely openwork weaves. In the experimental part of the work the methods of mathematical planning and analysis of results were used. Processing of experimental results was carried out by the method of mathematical statistics. In the technological part the modern weaving machine "Super Excel" of "Somet" company, devices of CENTEXUZ TITLP certification centre were used.

Scientific novelty of the work. In carrying out the theoretical, experimental and technological studies by the authors:

- dependences of warp and weft fabric strength, warp and weft bending will heights on the ratio of yarn diameters in falsely openwork weave fabrics were determined;
- Limit fabric densities for warp and weft of falsely openwork weave fabrics are revealed;
- An analytical calculation of warp and weft yarns processing is proposed, taking into account the number of transitions of each yarn instead of average transitions;
- falsely openwork weave fabrics have been designed and developed;
- The influence of the type and linear density of the weft, warp and weft

tension on the workmanship and production tension of falsely openwork weaves was investigated;

- defined physical and mechanical properties of falsely openwork fabrics weaves.

Practical value of the work. Samples of fabrics of falsely openwork weaves, which have a good appearance, good air permeability, have a rare combination of aesthetic-hygienic and physical-mechanical properties, that can be recommended the use of these fabrics for clothing. The technology of production and technical calculation of falsely openwork fabrics on the basis of local raw materials (cotton) is developed, which will allow to introduce the results of work in the industry and to solve the problem of shoe fabrics production of falsely openwork weaves, which are in great demand in the republic. Besides, the method of calculation of yarn processing can be used in the educational process when studying the structure and design of fabrics.

Content of work

The introduction substantiates the relevance of the topic, formulates the purpose, objectives of the study, shows the scientific novelty and practical significance of the work, methodology, research and approbation of the work.

In the first chapter a review of literature sources devoted to the design and mechanisms of formation of openwork and falsely openwork weaves is given. The expediency of obtaining fabrics imitating openwork weaves is substantiated. The aim and tasks of the research are shown. Fabrics of leno weave have higher resistance to sliding in comparison with plain weave fabrics. The reason for this is the greater number of crossing points of the yarns and the larger angles of coverage of the yarns by each other. If plain and leno weaves have the same yarn density, yarn tension and coefficient of friction, the sliding resistance of the leno weave is 1.7 times higher.

This enables a 30 % reduction in material consumption and a 40 % increase in productivity. The colours of the weft yarns dominate and the colours of the warp are hidden, which makes it possible to produce different fabrics with the same warp by changing the weft or changing the weft effect. The leno weave technique makes optimum use of the advantages such as gentle yarn handling, user-friendliness, cleanliness, ease of mastery as well as fast article changes. Thanks to two different tensions, a new type of fabric is formed. The warp yarns are crossed on the wrong side to increase the density of the weft, and thanks to the change in geometry, the fabric has an interesting new appearance. By replacing the interlocking yarns with fine yarns, it is possible to develop fabric constructions with the same properties on both sides. The filling threads can be so close together that the density factor can reach 100 %. The front side looks like a plain weave fabric. There is also an increase in strength of more than 20 per cent and the elongation characteristics are significantly improved. Advantages: firstly, new double-sided fabrics; secondly, reduced elongation of the fabric (weft and warp yarns are arranged in a straight line); thirdly, a higher degree of density and bright colours; fourthly, a relief surface due to the use of yarns with different linear densities; fifthly, the adjustment of different densities allows to obtain a mesh structure and high density stripes. It allows the production of fabrics with variable densities, because the weft is not processed due to the construction of the fabric. For all density settings, one twist roller with two needle rings is sufficient to guide the fabric. For the production of all types of leno weave fabrics - from light curtains to the heaviest fabrics. It allows the development of new fabrics for domestic and technical purposes. Interesting solutions are offered in the field of

carpet backing production

as well as a new textile floor covering which can be produced in one technological process. The production of leno weave fabrics is becoming more flexible, curtain manufacturers can produce leno and plain weave fabrics on the same walls. The artistic design is almost limitless - decorative and garment fabrics with longitudinal stripes of warp yarns in plain weave are possible. Clothing: surfaces like plain and sateen weave fabrics with a certain breathability in a variety of colours of high saturation. Combinations can be on a standard basis, exclusively through colour selection and variable weft density, e.g. for shirt fabrics, decorative fabrics. Reduced overall material consumption for furniture fabrics compared to classic sateen weave fabrics with better covering power and the same tear resistance. Also suitable for curtain fabrics due to the high colour brightness. Furniture fabrics: non-slippery surface embossed for better air circulation due to the different linear density of the warp yarns. Good deformability due to elastic yarns in the warp. Technical textiles. Constant warp density across the entire width of the fabric variable weft density coarse or fine meshes high strength are decisive factors when producing fabrics for tent coverings and all types of glass fabrics. Carpets. Velour surface structure with optimum hardening lightweight, environmentally friendly and produced in a single production process. From conventional fabrics to leno weave fabrics. The example of a furniture fabric shows the economic aspects and compares the properties of a primary leno fabric with those of a conventional fabric. In this case the productivity increase reaches 125 % and the material consumption is reduced by 28 % with the same or better product properties. Fabrics that have an openwork effect on the surface, obtained by lacing warp yarns of one system with yarns of another system are called leno or openwork. Openwork effects can be located on the fabric across the entire width of the fabric and as in the form of individual stripes, squares, checkers, etc. The peculiarity is that during the weaving process, one warp system (leno) is wound on the right and left relative to the other warp system (stockinette). On the weaving machine, the stock yarn **C** has a filling tension two to three times higher than the leno yarn **P**. This is in order to obtain a clear, clean and equal shed during shedding for each revolution of the main weaving machine. The formation of the leno weave on the fabric is shown in Fig. 1.1.

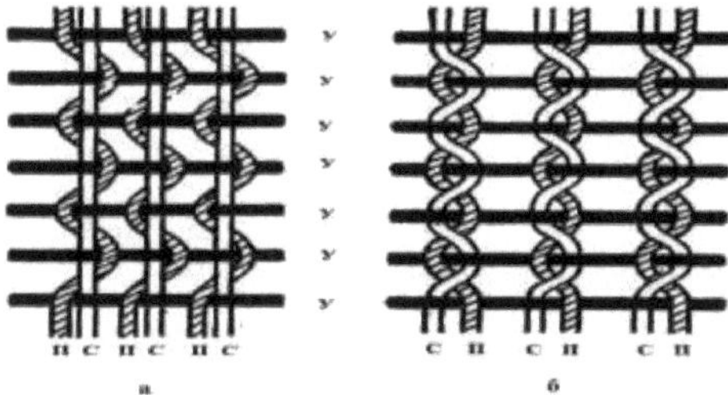

Fig. 1.1 Location of the **C** and **P** warp threads in the fabric, **a - location of** the **C** and **P** warp threads in the fabric directly on the loom, **b** - location of the **C** and **P** warp threads in the fabric after removal from the loom. **C- stocking** thread, **P leno** thread, **U filling** thread.

On the weaving machine, the filling threads **C** have a high filling tension and the leno threads **P** a minimum tension, so it is advisable to use a means to adjust the tension of the filling thread **C** and the leno thread **P** independently of each other. Wrapping of the stock thread with leno threads is carried out by special leno devices. Let's consider some varieties of lacing devices.

The first leno device is shown in Fig. 2.4, where the loops 1, 2 and Z of the leno frames are connected to the eye G, in which the leno thread P is inserted.

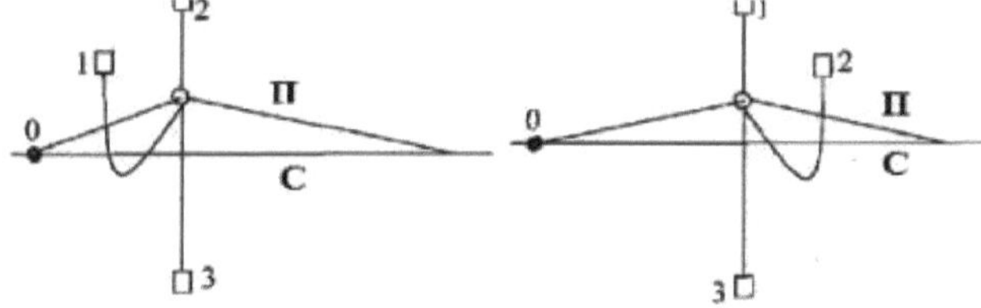

First pharynxSecond pharynx

Fig. 1.2. Lacing attachment on the weaving machine.

In the first weft cast-on (first shed), loop 2 is raised and the leno thread **P is** raised together with it on the left side of the warp **C,** while loop 1 is lowered. With the second filling thread (second shed), loop 1 is lifted, and with it the leno thread **P** on the right side of the stockinette warp **C**, and loop 2 is lowered. Loop 3 serves for the correct movement of loops 1 and 2. In the second type of devices (Fig. 1.2), the stocking thread **C** sneaks into the eye of the galeva of the stocking remizki 1, and the flexible galevo (loops) with the eye G carrying the leno thread and covers the stocking thread **C** and is connected to the leno remizki 2 and 3. At the first weft threading (the first shed), leno 2 moving upwards will position the loop with the eye and together with them the leno thread P on the left side of the stock thread C. The filling thread laid in the shed and nailed to the fabric edge will fix the leno and stocking warp threads in this position. At the second revolution of the main shaft (shed change), the thread looper 3, moving upwards, will position the eyelet with the loop and, as a consequence, the leno thread to the right of the warp thread C.

Fig. 1.4 shows the third type of ligatures

A ligature pair, where the stock yarns **C are** threaded through the eyes of the halewing harness 1, and the leno yarns **P** through the eyes of the halewing harness 2 and into a leno pair consisting of a wing 3 and a half wing 4. Wing 5 is an ordinary remizka in the eye of the halev, which is pierced with flexible loops (of twisted thread or monofilament) of the half-wing. At the first shedding, the leno thread **P is** positioned to the left of the stock thread **C** due to the lifting of the leno thread 2 and the wing 4, while the wing 3 and the stock thread 1 are lowered.

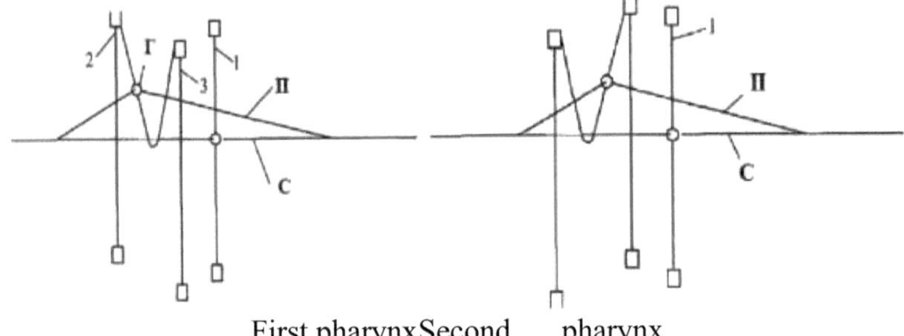

First pharynxSecond pharynx

Fig. 1.3. Lacing attachments on the weaving machine.

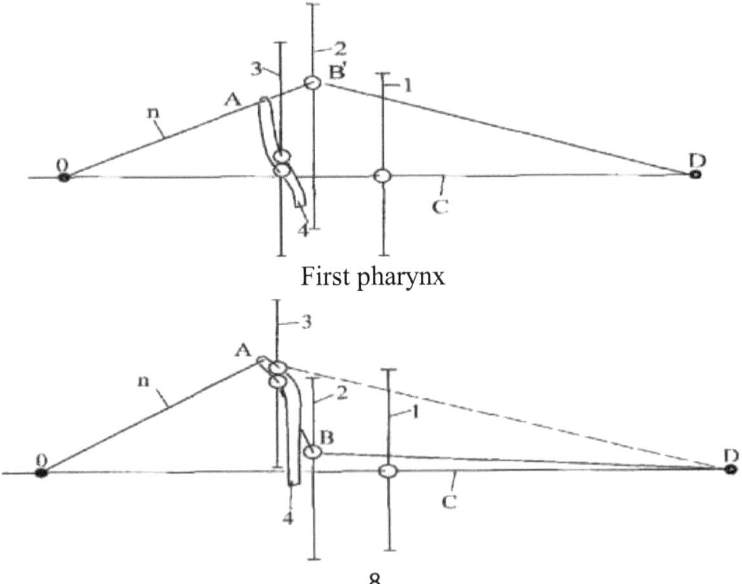

Second pharynx

Fig. 1.4. Lacing attachments on the weaving machine.

In the second shed (second sinking), the leno thread **P is** positioned to the right of the stock thread **C** by lifting the wing 3 and the half wing 4, while the leno threader 2 and the stock threader 1 are lowered. As can be seen, at each revolution of the main shaft of the machine, the half wing is in the upper position, and the haystring 1 in the lower position. The formation of the first shed does not cause a high tension of the leno threads P, because the leno loom 2 and the half wing 4 are at the top with the geometry of the leno thread filling OAD. The formation of the second shed causes a large amount of tension on the leno threads, as wing 3 and half wing 4 are in the up position and leno loom 2 is in the down position, which changes the shed filling geometry of the OAD. The following methods are used to compensate for a leno length equal to (AB+BD-AD):

1. The installation of an additional forced - oscillating scalo for the leno warp, which loosens the leno warp when a second shed is formed.
2. Turning the leno with the leno base by a small additional angle when the second shed is formed.
3. Installation of an additional (compensating) heddle, in which the leno threads are inserted into the galeva, which loosen the leno threads of the warp when a second shed is formed. When producing leno fabrics, except for flexible (threaded) galleys, rigid (metal, plastic) galleys are used. Fig. 1.5 shows a metal galeva with an eye in the upper part of the galeva, in which the stocking thread **C is** threaded, with the outer surface of the galeva in contact with the leno thread **P**. Lowering the left wing 2 leads to downward movement of the leno thread **P** and its location to the left of the stocking thread **C**. And lowering the right wing 3 leads to downward movement of the leno thread **P** and its location to the right of the stocking thread **C.**

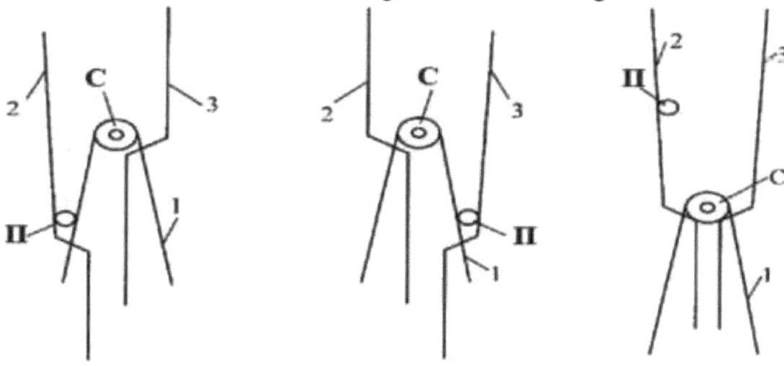

Fig. 1.5 The leno devices on the weaving machine.

Rigid galleys occupy more space than flexible galleys (threaded galleys), in the process of leno formation, leno threads may be clamped, which leads to increased thread breakage. In this paper, a system for producing leno weave for fabric production on modern weaving machines has been developed. Fig. 1.6 shows a schematic diagram of the mechanism and Fig. 1.7 shows the position of the slats in different phases. On the shaft 1 are mounted cams 2, contacting the rollers located on a three-arm lever 4. The levers 4 are connected to the ligature laths 10 (101) through links 5, two-shoulder levers 6 (61), rods 7, angle levers 8, rods 9. Each galewo 11, carrying a leno thread 12 in the eyes, is connected by its ends to the upper galewo carriers of the leno remises 10 and 101. The mullion thread 13 passes over the galev 12 and is threaded through the eye of the galev 14 by the mullion remise 15. The rack 15 is respectively connected to the double-shoulder arm 16 via the rods 9, angular levers 8, rod 7. The means for moving the packer 15 in the scoring position is in the form of a pin 17 and a spring 18. The finger 17 is rigidly seated on the lever 16 and is in contact with the shoulders of the levers 6 (61) of the ligature bars 10 (101). The spring 18 is connected at one end to the pin 17 and at the other end is fixed to the loom frame. By repositioning the pin 17 in the hole 19 of the lever 16, the amount of movement of the packer slats 15 from the line of scoring to the lowest position is changed. In addition, the pin 17 is mounted eccentrically on the lever 16 (not shown in Fig.). This makes it possible to adjust the position of the pack yarns relative to the weft weaver during its movement through the shed. The length of the vertical arm of the lever 4 regulates the amount of movement of the leno-strands 10 and 101. Yaw-forming mechanism works as follows. Cams 2 rotating, transmit motion through levers 4, rods 5, levers 6 (61), rods 7, levers 8, rods 9 to ligature lashings 10 (101). The linting laths 10 and 101 operate in a plain weave mode. The movement of the pack lath 15 is received from the levers 6 (61) through the pin 17, spring 8, lever 16 and the corresponding links 7, 8, 9. The character of the movement of the foot harrow 15 is as follows - initial (lower) position, movement to the line of scoring and movement to the initial (lower) position. Up to the line of scoring, the leno lath 101 leads the stock lath 15 (position 1 in Fig.) by turning the lever 61 clockwise, in the same direction the lever 16 is turned through the pin 17 and the spring 18. In position II, the eyes of the gall eyes 11 and 14 of the lashings 10, 101 and 15 are on the line of engagement, respectively at this moment are in the same plane of the lever 6, 61 and 16. Continuing its clockwise movement, the lever 61 will move away from the pin 17. At the same time, lever 6 rotating anti-clockwise leads lever 16 in the same direction through pin 17, at the same time spring 18 is stretched. As a result, the slats 10 and 15 are placed in the lower position (position III). To form openwork fabrics, the leno threads 12 are arranged to the left (position I) and right (position III) relative to the stock thread 13. By moving the stocking threader 15 to the scoring position (position II), the conditions for the transition of the eyes of the leno threads 12 relative to the stocking threads 13 are improved. In addition, the deformation of the leno threads 12 is reduced, due to the reduction of the movement of the leno bars 10 and 101 (reduction of the shed height). Also, the proposed mechanism is easy to maintain and simple in design, as it excludes the options of using an additional drive for the stock loom. To produce leno fabrics with no more than 12 threads per 1 cm, special machines are used, in which the stock and leno galleys (remizki) are replaced by bars with needles, making a lateral movement to form a leno, and then in height to form a shed. Fabrics of leno weaves are produced from threads and yarns of different types of linear density, combinations of type and colour. They are used for making dresses, blouses, curtains, sieves, filters and other products, as well as leno-weaves weaves are used to secure the fabric edges on weaving machines.

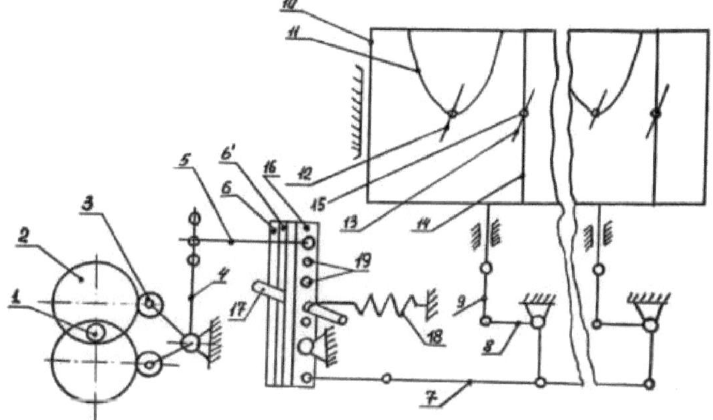

Fig. 1.6. Lacing attachments on the weaving machine.

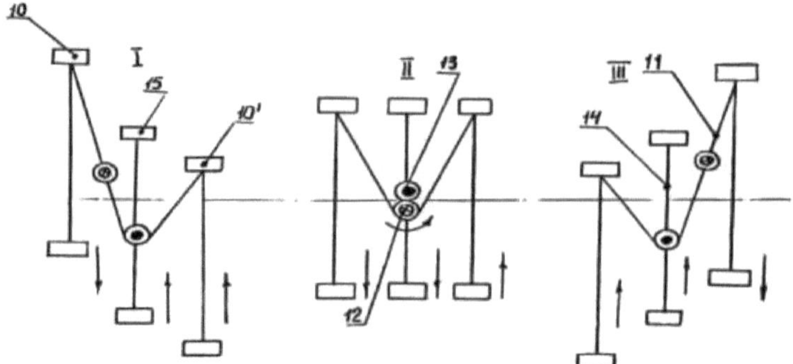

Fig. 1.7. Position of the remise in different phases.

Fabrics of leno weaves require special equipment (weaving device parts) dressing (additional filling) and compensating mechanisms (tension and leno thread supply), etc., which causes difficulties in their production. Therefore, it is expedient to produce such fabrics on weaving machines by imitating openwork (leno) weaves at the expense of falsely openwork (falsely leno) weaves. False leno fabrics, being the simplest in terms of the method of production and without any lacing of the main threads, give the impression of leno. There are two types of false leno weaves: false smooth leno weaves; false figured leno weaves.

The first imitates smooth leno weaves, creating through-holes in the leno areas. The second imitates shaped leno weaves, which create embroidery-like patterns in the form of main threads zigzagging on the front surface of the fabric.

the use of special weaves. For those main yarns in false figure leno fabrics, which create the effect, cotton yarns are mostly used, such as stark, bleached and coloured yarns, twisted in several low-numbered threads.

False smooth leno weaves. These fabrics have through-through translucent openings because certain groups of main and weft yarns are pulled apart from each other to form longitudinal and transverse openings between them. The sliding of these groups of threads is due to the fact that the outermost main and outermost weft threads of each group have an opposite weave. As a consequence, when interlocking the differently interwoven main threads of each group, the groups are pushed apart by the weft threads in the longitudinal direction; when interlocking the differently interwoven weft threads of each group, the groups are pushed apart by the main threads in the transverse direction; together with this, the main threads in the transverse direction; together with this, the main and weft threads of each group, as a result of their interlacing, are easily shifted close to each other; the interlacing of the threads of each group as a whole - completely unlike the interlacing of the main and weft threads of each group. Many false leno weaves are constructed according to the principle of combining the weave of corduroy with the fabric; therefore, due to the established fact that fabrics with corduroy weave need a higher density compared to fabrics of other weaves, the gaps between groups of threads are increased as a result of insufficient density of corduroy elements in fabrics of false leno weaves. The imitation of shaped leno weaves with effects in the form of wavy main threads is achieved by the fact that one part of the weave pattern with a large number of crossing points (almost always a cloth) alternates with another part of the weave pattern with loosely arranged threads - without interlacing. As a consequence, the corresponding main threads, having a plain weave on one side and a free weft on the other, tend to move in the direction of the weft, being arranged in a wavy manner on this section of the fabric. False figurative leno weaves are also achieved by using the main threads twisted in several threads with a large coefficient of twist, and ordinary (only spun); in this case, those and other threads have a different direction of twist. For example, along with ordinary single threads of right twist there are twisted threads of left twist, remaining unintertwined for a long distance, as a result of which they take different positions on the surface of the fabric.

Chapter 2 deals with the structure of falsely openwork weaves and factors determining the structure of fabrics. The structure of a fabric is usually understood as the mutual arrangement of warp and weft yarns in the fabric due to their interaction. The forces of interaction between the threads in the fabric are created in the process of its formation on the loom and determine the mutual arrangement of threads in the fabric. The mutual arrangement of the threads in the fabric depends on many factors: the type of raw material used; the diameter of the warp and weft threads and their ratio; the warp and weft densities and their ratio; the type of weave in the fabric; the tension of the warp and weft threads and the tension ratio; the technological parameters

dressing and production of fabrics. The type of raw materials for the designed fabric is selected taking into account the purpose of the fabric and the requirements to it. The properties of the yarns used in warp and weft largely determine the properties of the fabric made from them. A change in the type of raw material in at least one system of yarns in the warp or weft of the fabric has a significant impact on the technological parameters of its production, the structure of the fabric and its properties. The diameters of warp and weft yarns used for fabric production have a significant influence on the technological parameters of fabric production, on its structure and properties. When designing a fabric, yarn diameters are determined depending on the purpose of the fabric and the requirements to it. Increasing the diameter of weft yarns increases the breaking

load and elongation of the fabric in the weft direction, the working out of the main yarns and reduces the working out of the weft. Consequently, the ratio of warp and weft yarn diameters has a great influence on the parameters, structure and properties of the fabric. The warp and weft densities and their ratios have a great influence on the structure and properties of fabrics. The change of fabric density by weft, other things being equal, causes the change of technological parameters of production, structure and properties of fabric. In particular, an increase in the weft density of the fabric leads to an increase in warp tension and yarn workmanship.

decrease of warp yarns processing and fabric width.

The warp and weft densities depend on the diameter of the yarns used and the type of weave of the yarns in the fabric. The maximum possible density of plain weave fabrics is lower than any other weave in the fabric (corduroy, rayon, twill, sateen, etc.). Fabrics with the highest possible warp and weft densities are very difficult to weave on the loom. Most of the fabrics produced have a density of one or both yarn systems that is less than the maximum. Therefore, the ratio of the actual fabric density to the maximum density characterises the filling of the fabric with fibrous material, i.e. the tension of the fabric production on the loom. The type of weave has a great influence on the structure and properties of the fabric. In particular, plain weave fabrics have a higher breaking load and warp and weft yarn working than fabrics of other types of weaves (corduroy, rep, twill, sateen, etc.), and the working of the fabric (plain weave) is accompanied by a higher tension. Among the technological parameters that have a significant influence on the structure and properties of the fabric are the tension of the main and weft threads and their ratios, which change the arrangement of threads in the fabric, and therefore the working of threads in the fabric, the tensile load of the fabric. Another main parameter of machine dressing is the value of the backstop, which determines the value of additional warp tension at the moment of fabric formation (surf). As the backstitch increases, the tension of warp threads at the moment of surf increases, which leads to a change in the structure of the fabric, in the density to a change in the workmanship of threads in the fabric, tensile load and elongation of the fabric. Hence, with a change in the filling tension and the amount of the machine scoring, the structure and properties of the fabrics can be changed. In addition, the structure and properties of fabrics are influenced by the position of the scalo, the height and depth of the shed, the position of the spar, etc. Consequently, all the above parameters together determine the structure of a fabric with the corresponding arrangement of threads of both systems in it. The structure of the fabric also determines its physical and mechanical properties: strength, elongation, stiffness, drapeability, hygroscopic, heat-protective and other properties. The mutual arrangement of warp and weft yarns in a fabric is determined by their bending, namely, the height of warp (h_o) and weft ($\boldsymbol{h_y}$) bending wills and, respectively, the length of their half-wave (l_o and l_y). The bend wave height is the distance between the levels of the yarns of the same system in the vertical plane at warp and weft overlap. The half-wave and weft length in a single ply fabric or in one ply of a multilayer fabric is determined by the corresponding distance between two neighbouring yarns of the opposite system in the system at their intersections.

One of the main questions of fabric structure is to determine the height of bending waves of warp and weft yarns in the fabric and to find the analytical dependence between wave heights, half-wave lengths, warp and weft yarn diameters, as well as fabric density for both yarn systems, which will allow us to approach the engineering design of fabrics. Such analytical dependence for studying the structure of the fabric is carried out in the works, where a number of assumptions are made, the fabric of plain weave, in which warp and weft threads have equal diameters: $do = dy = 2r$, i.e., $do : d_y = 1:1$, the threads in the fabric retain a cylindrical shape and *do* not change their dimensions in the process of forming the fabric on the loom. Fig. 2.1 shows a diagram of the mutual arrangement of warp and weft yarns at different orders of construction phase.

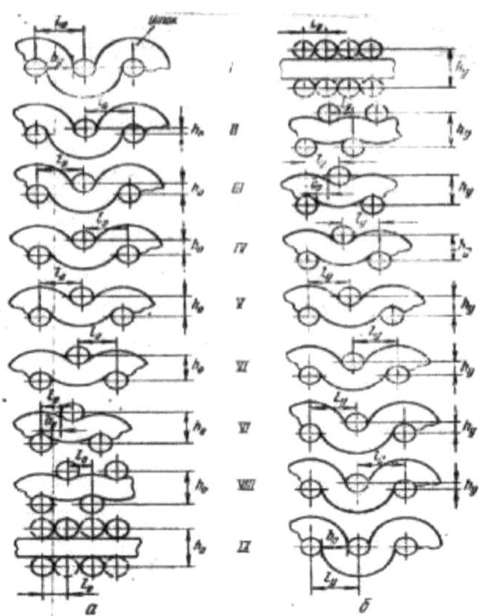

Fig. 2.1. Schemes of the mutual arrangement of warp and weft yarns at different orders of construction phase: a - section along the weft yarn; b - section along the warp yarns

From Fig. 2.1 it follows that the main threads are arranged in the fabric in a straight line, and the weft threads bend around the main threads, The height of the weft bending wave is maximum $hy = max$, and the height of the warp bending wave is 0 (see Fig. 2.1 position *I*), The weft threads are arranged in the fabric in a straight line, and the main threads bend around the weft threads and have maximum bending. i.e. $\boldsymbol{h_o} = max$, $\boldsymbol{h_y} = 0$ (see Fig. 2.1 position *IX*). At ($do=dy$), each order of the construction phase differs from the previous and subsequent order by shifting the yarns in the vertical plane by half of the yarn radius. For example, at the first order of construction phase $ho=0$, $hy=4r$, and at the second order of construction phase $ho=0.5r$, $hy=3.5r$. The fabric of the fifth order of structure phase has $ho=hy=2r$. The basic geometrical property of a single-layer fabric: the sum of heights of bending waves of warp and weft yarns in the fabric is a constant value equal to the sum of yarn diameters, i.e., the sum of thread diameters.

$$ho + hy = do + dy = const$$

If the bending wave height of the filaments of one system decreases, the bending wave height of the filaments of the other system increases by the same amount.

The distance between the centres of two neighbouring yarns (lo and ly) in plain weave determines the geometric density of the fabric, which is the inverse of the technological density of the warp and weft (P_o and P_y). By its definition, the geometric density is identical to the half-wavelength of the bend of the yarn of the opposite system. Hence, for plain weave fabric: $l_o = 100 : P_o$ and $l_y = 100 : P_y$. As can be seen from the scheme shown in Fig. 2.1, with the change of the fabric structure phase order from I to IX at the maximum possible technological density, the distance between warp yarns, i.e. the geometric density of the warp (l_o) decreases, while the distance between weft yarns, i.e. the geometric density of the weft (l_y), increases. Consequently, under all other conditions being equal and $d_o=d_y$, fabrics from I to IV order of the construction phase have a higher technological density in weft than in warp. For the fabric of the V order of the construction phase $P_o=P_y$, and for the other orders of the construction phase the fabric should have a greater technological density in the weft than in the warp. To analyse the fabric structure of any weave having a different ratio of warp and weft diameters, it is necessary to derive the basic geometrical property of fabric structure through the average yarn diameter

$h_o + h_y - d_o + d_y = 2d_{cp}$, *(2.1)* where d_o and d_y *are the* actual thread and fabric diameters;

d_{cp} - average actual thread diameter in the fabric;

$$d_o + d_y / 2 = d_{cp} ,$$

In this case, the accepted definition of the change of the fabric structure phase order through the thread radius is replaced by its average diameter. Depending on the order of the fabric structure phase, the heights of the thread bending waves that retain the cylindrical shape in the fabric can be determined by the formulas:

$$h_o = K_{ho} d_{cp}, \quad (2.2)$$

$$h_y = K_{hy} d_{cp} , \quad (2.3)$$

where; K_{ho} and K_{hy} are the coefficients that take into account the change in the height of bending waves of warp and weft yarns in the fabric depending on the order of the phase of its structure,

Substituting the values of $\boldsymbol{h_o}$ and $\boldsymbol{h_y}$ into formula *(2.1)* we have:

$$K_{ho}\, _{cp}d + K_{hy}\, _{cp}\, _{cp}d = 2d,$$

Hence we obtain the basic position that the sum of the coefficients determining the heights of the thread bending waves for any order of the fabric structure phase is a constant value equal to two: $K_{ho} + K_{hy} = 2$. When the structure of the fabric changes by one order, the coefficients change by 0.25; the value of one coefficient decreases by 0.25 and the value of the other increases by the same amount. For the first order of structure phase $\boldsymbol{K_{ho}} = 0,$

and K_{hy} = 2, respectively and $h_o = 0$, $h_y = 2d_{cp}$, For the *V* order of structure phase $K_{ho} = K_{hy} = 1$ and $h_o = h_y = d_{cp}$. For IX order $K_{ho} = 2$, $K_{ho} = 0$ and $K_{hy} = 0$ and $h_o = 2d_{cp}$, $h_y = 0$. When moving from one order to the next, the height of the warp bending wave increases by 0.25 d_{cp} , and the height of the weft bending wave decreases by the same amount, The ratio of the yarn bending wave heights is expressed as follows

by the coefficient $K_h = K_{ho} : K_{hy} = h_o : h_y$, (2.4)

Table 2.1 shows the data showing the change of K_{ho} and K_{hy} coefficients, warp and weft bending wave heights at the change of fabric structure phase,

Table 2.1.

Variation of coefficients and wave heights of warp and weft bending yarns under changing phase of fabric structure,

Phase order	K_{ho}	K_{hy}	h_o	h_y	Bending wave height ratio coefficient K_h
I	0	2	0	$2d_{cp}$	0
II	0,25	1,75	0.25 d_{cp}	1.75 d_{cp}	0,143
III	0,5	1,5	0.5 d_{cp}	$1{,}5\,d_{cp}$	0,333
IV	0,75	1,25	0.75 d_{cp}	1.25 d_{cp}	0,6
V	1,0	1,0	d_{cp}	d_{cp}	1,0
VI	1,25	0,75	1.25 d_{cp}	0.75 d_{cp}	1,66
VII	1,5	0,5	1.5 d_{cp}	0.5 d_{cp}	3,0
VIII	1,75	0,25	1.75 d_{cp}	0.25 d_{cp}	7,0
IX	2	0	$2d_{cp}$	0	yes

For a maximally compact plain weave fabric and any order of construction phase, the distance between neighbouring warp and weft yarns (geometric density of the fabric) determines the half-wavelength of the bending yarn of the opposite system:

$$l_o = \sqrt{(do + dy)^2 - h_o^2} = \sqrt{4d_{cp}^2 - K_{ho}^2 d_{cp}^2} = d_{cp}\sqrt{4 - K_{ho}^2},$$

$$l_y = \sqrt{(do + dy)^2 - h_y^2} = \sqrt{4d_{cp}^2 - K_{hy}^2 d_{cp}^2} = d_{cp}\sqrt{4 - K_{hy}^2}.$$

Coefficients determining the bending wave heights of warp and weft yarns in the tissue, characterise the order of the phase of its structure. The coefficients can have any intermediate value. The numerical value of the structure phase order can be expressed through the coefficients of bending wave heights:

$F = 4\ K_{ho} + 1$ or $F = 9 - 4\ K_{ho}$, Since $K_{ho}\ K_{hy} = K_h + K_{hy} = 2$,

then $K_{ho} = 2\ K_h / K_h + 1$, $K_{hy} = 2 / K_h + 1$, $K_{hy} = 2 / K_h + 1$- Substituting these values into the formula for determining the numerical value of the order of the phase of the tissue structure, we obtain a formula defining this value through the coefficient of the ratio of bending wave heights: $F = 9\ K_h + 1 / K_h + 1$

The same formulas allow us to express the coefficients K_{ho} , K_{hy} through the numerical values of the order of the tissue structure phase:

$K_{ho} = F - 1 / 4$, $K_{hy} = 9 - F / 4$, $K_h = F - 1 / 9 - F$.

To determine the phase order of a plain weave fabric, it is necessary to first find the geometric density of the warp or weft: $l_o = 100 : P_o$ and $l_y = 100 : P_y$. The height of the warp or weft bending wave in the fabric can be determined by solving the equations concerning

ho or ***hy*** : $h_o = \sqrt{4d_{cp}^2 - l_o^2}$ или $h_y = \sqrt{4d_{cp}^2 - l_y^2}$.

If the height of the warp wave has already been determined, we find the height of the weft wave: $h_y = 2d_{cr} - h_o$. When the height of the weft bending wave has been determined, we find the height of the warp bending wave: $h_o = 2\ d_{cf} - h_y$. To determine the phase order of any other weave it is necessary to find the average geometric density of the fabric on the warp and weft: $l_{osr} = 100 : P_o = L_{Ro} : R_o$, $l_{usr} = 100 : P_y = L_{Ry} : R_y$, where L_{Ro} and L_{Ry} *are the* length of the fabric within the weave rapport on the warp and weft; R_o and R_y *are the* weave rapport of the fabric on the warp and weft. Fig. 2.2 shows the layout of warp yarns in plain weave and plain weave.

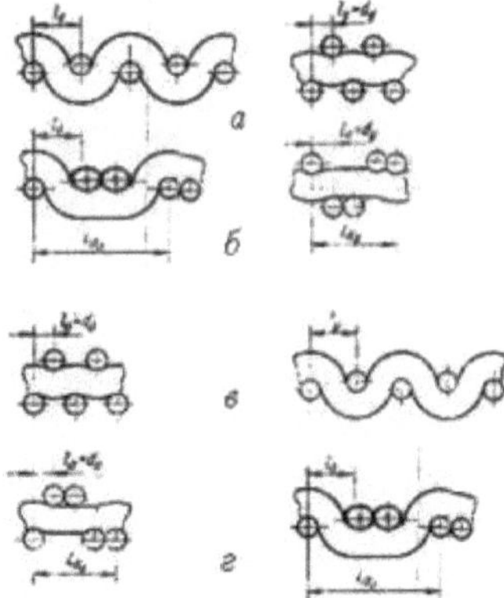

Fig. 2.2. Layouts of warp yarns in plain weave (a and c) and plain weave (b and d).

In the long overlaps of a maximally compacted fabric, the yarns of the opposite system are placed next to each other and the distance between neighbouring yarns is equal to the yarn diameter of the fabric. The length of the fabric within the warp and weft weave ratio can be determined by the formulas: $L_{Ro}=t_y\, l_o+(R_o - t_y)\, d_o$, $L_{Ry}=t_o\, l_y+(R_y - t_o)\, d_y$, where t_o и t_y *are the* number of weft to warp and warp to weft intersections. The average distance between warp and weft yarns (average geometric density) of the fabric is determined by the formulas: $l_{ocp} = t_y\, l_o+(R_o - t_y)\, d_o / R_o$, (2.5), $l_{ycp}= t_o\, l_y+(R_y - t_o)\, d_y / R_y$. (2.6). The distance between warp and weft yarns where they intersect with yarns of the opposite system can be determined by solving these equations with respect to l_o and l_y :

$$l_o = \frac{l_{ocp} R_o - (R_o - t_y)\, d_o}{t_y} = \frac{\frac{100}{P_o} R_o - (R_o - t_y)\, d_o}{t_y},$$

$$l_y = \frac{l_{ycp} R_y - (R_y - t_o)\, d_y}{t_o} = \frac{\frac{100}{P_y} R_y - (R_y - t_o)\, d_y}{t_o}.$$

Once t_o and t_y have been determined, the order of the structure phase can be found by analogy with plain weave fabrics.

In order to reveal the influence of the ratio of thread diameters on their arrangement in the fabric, let us consider the bending of threads in plain weave fabric. For the fabric when the diameter ratio is varied at $l_o = d_o$ and the fabric when the diameter ratio is varied at $l_y = d_y$.

The diameter of the warp threads in the fabric will be determined by the formula:

$$do = 2Kd\, dcp / Kd+1 \quad (2.6)$$

The diameter of the weft yarns in the fabric will be determined by the formula:

$$dy=2dcp /(Kd+1) \quad (2.7)$$

The ultimate density of the fabric on the base is determined by the formula

$$Po=100 : do. \quad (2.8)$$

The fabric density limit for the weft is determined by the formula

$$Py = 100 : dy. \quad (2.9)$$

The bending wave height of the warp yarns in the fabric at $l_o = d_o$ is determined by the formula:

$h_o = \sqrt{(do + dy)^2 - d_o^2} = 2d_{cp}\sqrt{1 - \frac{K_d^2}{(Kd+1)^2}}$. (2.10)

and the duck $h_y = 2d_{cp} - h_o$ (2.11).

Geometric density in weft $l_y = \sqrt{4d_{cp}^2 - h_y^2}$ (2.12)

Maximum weft density $P_{ymax} = 100 : l_y$ (2.13)

The bending wave height of the warp yarns in the fabric at $lu = d_y$ is determined by the formula:

$ho = 2dcp - hy$ (2.14), and by weft $h_y = 2d_{cp}\sqrt{1 - \frac{K_d^2}{(Kd+1)^2}}$. (2.15)

Geometric density on the base $l_o = \sqrt{4d_{cp}^2 - h_o^2}$ (2.16)

Maximum basis density $Romah = 100 : l_o$ (2.17)

The above-mentioned parameter values determine the structure of falsely openwork fabrics, which have plain weave, rape weave, corduroy weave or planking overlaps in the weave. The effect of weave, warp and weft yarn diameter and their ratio on the structure of fabrics is also examined. On the basis of changing the coefficient of the ratio of yarn diameters from 0.5 to 2 for linear density of warp and weft yarns $T_o = T_y$ =20 tex, the diameters, height of bending waves, geometric density of yarns, limit and maximum density of fabric, and coefficients determining the order of the phase of fabric structure were calculated. Table 2.2 presents the results of calculation of fabric structure parameters at change of diameter ratio coefficient at $lo = do$, and Table 2.3 presents the results of calculation of fabric structure parameters at change of diameter ratio coefficient at $lu = d_y$.

Table 2.2.

Results of calculation of tissue structure parameters at changing the diameter ratio coefficient at $lo = do$.

Diameter ratio Kd	Diameter of warp yarns do, mm	Diameter of weft yarns dy, mm	Ultimate density on base Po, n/dm	Height of bending waves		Geometric density in weft lu, mm	Maximum weft density $Rumah$ n/dm	Phase order factor, Kho
				ho, mm	duck, mm			
0,5	0,109	0,217	917,4	0,307	0,019	0,3254	307,3	1,88
0,6	0,122	0,204	819,7	0,302	0,024	0,3251	307,6	1,85
0,7	0,134	0,192	746,3	0,297	0,029	0,3247	308,0	1,82
0,8	0,145	0,181	689,6	0,292	0,034	0,3242	308,5	1,79

0,9	0,154	0,172	649,4	0,287	0,039	0,3237	308,9	1,76
1,0	0,163	0,163	613,5	0,282	0,044	0,3230	309,6	1,73
1,2	0,178	0,148	558,7	0,273	0,053	0,3217	310,8	1,68
1,4	0,190	0,136	526,3	0,265	0,061	0,3202	312,3	1,63
1,6	0,201	0,125	497,5	0,257	0,069	0,3186	313,9	1,58
1,8	0,210	0,116	476,2	0,249	0,077	0,3168	315,7	1,53
2,0	0,217	0,109	460,8	0,243	0,083	0,3153	317,2	1,49

Table 2.3.

Results of calculation of fabric structure parameters at changing the diameter ratio at $l_u = d_{\cdot y}$

Diameter ratio K_d	Filament diameter d_o, mm	Diameter of weft yarns d_y, mm	Ultimate duck density P_y, n/dm	Height of bending waves		Geometric density on the base l_o, mm	Maximum density by base P_{ohms}, н/дм	Phase order factor, $K_{|,y}$
				h_o, mm	duck, mm			
0,5	0,109	0,217	460,8	0,083	0,243	0,3153	317,2	1,49
0,6	0,122	0,204	476,2	0,077	0,249	0,3168	315,7	1,53
0,7	0,134	0,192	497,5	0,069	0,257	0,3186	313,9	1,58
0,8	0,145	0,181	526,3	0,061	0,265	0,3202	312,3	1,63
0,9	0,154	0,172	558,7	0,053	0,273	0,3217	310,8	1,68
1,0	0,163	0,163	613,5	0,044	0,282	0,3230	305,6	1,73
1,2	0,178	0,148	649,4	0,039	0,287	0,3237	308,9	1,76
1,4	0,190	0,136	689,6	0,034	0,292	0,3242	308,5	1,79
1,6	0,201	0,125	746,3	0,029	0,297	0,3247	308,0	1,82
1,8	0,210	0,116	819,7	0,024	0,302	0,3251	307,6	1,85
2,0	0,217	0,109	917,4	0,019	0,307	0,3254	307,3	1,88

The graphs (Fig.2.3-2.4) of dependence of fabric density on warp and weft, height of bending waves of warp and weft yarns on the ratio of diameters, at geometrical density on warp equal to the diameter of the main thread and geometrical density on weft equal to the diameter of the weft thread have been constructed.

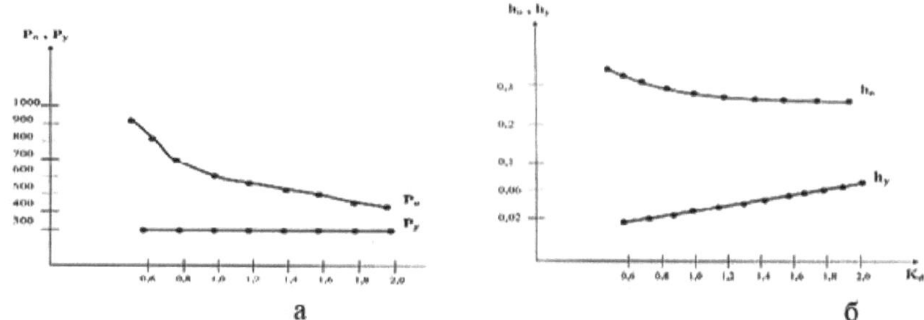

a б

Fig. 2.3. Graph of dependence of fabric density and height of bending waves of
warp and weft yarns
on the diameter ratio coefficient at, $l_o = d_o$

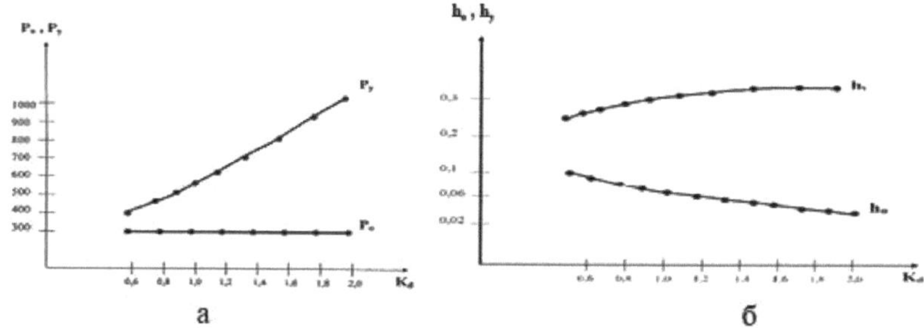

a б

Fig. 2.4. Graph of dependence of fabric density and height of bending waves of warp and weft yarns on the diameter ratio coefficient at, ***$l_y = d_y$***

The graphs show that when the ratio of yarn diameters is changed from 0.5 to 2: for geometric density on warp equal to the diameter of the warp yarn, the ultimate maximum density and bending wave height of the warp yarns significantly decrease, and the ultimate maximum density and bending wave height of the weft yarns slightly increase; for geometric density on weft equal to the diameter of the weft yarn, the ultimate maximum density and bending wave height of the warp yarns decrease, and the ultimate maximum density and bending wave height of the weft yarns increase. Hence, the ratio of bending waves (warp-weft) shows the formation of falsely openwork fabrics on the loom, i.e. for the first variant (Fig. 2.3) the fabric structure will have between the seventh and eighth phase of fabric structure as $h_o/h_y>1$ and for the second variant (Fig. 2.4) the fabric structure will have the second and third phase of fabric structure as $h_o/h_y<1$.

The influence of weave on the structure of plain weave and plain weave fabrics on the change of the maximum possible technological and minimum possible density and geometric density, the order of the structure phase, as well as the limiting orders of the structure phases is also investigated. The fabric is made of cotton yarns of thickness T_o = 20 tex, and T_y = 20 tex. In addition, the limiting maximum warp and weft densities were determined for plain weave and cambric weave. Table 2.4 shows the results of calculation of the structure parameters of plain weave fabric, and Table 2.5 shows the results of calculation of the structure parameters of plain weave fabric.

Table 2.4.

Calculation results of the structure parameters of plain weave fabrics

Tissue structure phase order	Coefficient determining the height of bending waves		Height of bending waves, mm.		Geometric density, mm		Maximum yarn density/dm	
	based on ***Kho***	on the duck ***Khy***	on the basis of ***ho***	by duck y^h	***lo***	on duck ***ly***	by ***Po*** basis	on duck ***Py***
Marginal	*0,27*	1,73	0,044	0,282	0,323	0,164	310	610
III	*0,5*	1,5	0,082	0,244	0,316	0,216	316	463
IV	*0,75*	1,25	0,122	0,204	0,302	0,254	311	394
V	1	1	0,163	0,163	0,282	0,282	355	355
VI	1,25	0,75	0,204	0,122	0,254	0,302	394	332
VII	1,5	0,5	0,245	0,081	0,215	0,316	465	316
Marginal	1,73	0,27	0,282	0,044	0,164	0,323	610	310

Table 2.5.

Results of calculation of parameters of structure of the fabric weave corduroy.

Tissue structure phase order	Coefficient determining the height of bending waves		Height of bending waves, mm.		Geometric density, mm		Maximum yarn density/dm	
	based on ***Kho***	on the duck ***Khy***	on the basis of ***ho***	by duck y^h	***lo***	by duck ***ly***	by ***Po*** basis	on duck ***Py***
Marginal	0,27	1,73	0,044	0,282	0,243	0,164	412	610
III	0,5	1,5	0,082	0,244	0,240	0,190	417	526
IV	0,75	1,25	0,122	0,204	0,232	0,209	431	478
V	1	1	0,163	0,163	0,222	0,222	450	450
VI	1,25	0,75	0,204	0,122	0,209	0,232	478	431
VII	1,5	0,5	0,245	0,081	0,189	0,240	329	417
Marginal	1,73	0,27	0,282	0,044	0,164	0,243	610	412

Comparison of the values reveals that for plain weave it is 32% higher than for plain weave. It is difficult to weave fabrics with the maximum possible density of both yarn systems on the loom, so we get fabrics with less than the maximum possible warp and weft densities. The filling factor is the ratio of the actual fabric density to the maximum possible density:

$Kno = Po/Po$ and $KHy = Py/Pu,$ where P_o and P_y are the actual technological density of the fabric in warp and weft;

P_o and Ru - the maximum possible warp and weft density of the fabric; l_{ocp} and l_{ycp} - *the* average distance between warp and weft yarns in the fabric;

$$locp\ LRo / Ro \text{ and } lycp\ LRy / Ry,$$

where L_{Ro} and L_{Ry} are the length of the weave rapport in the warp (along the weft yarns) and weft (along the main yarns), taking into account the order of the construction phase; R_o and R_y are the number of threads within the weave rapport.

In long overlaps of one system, the yarns of the other system are placed next to each other, as the free spaces between them are eliminated. In order to determine the length of the warp and weft rapport of the weave, the number of threads in the rapport and the number of overlaps (overlap changes) on each thread must be calculated from the weave pattern.

Consider the false openwork weave (Fig. 10) its basic parameters. Rape of the weave on the warp $Ro = 6$ threads, on the weft $Ry = 6$ threads.

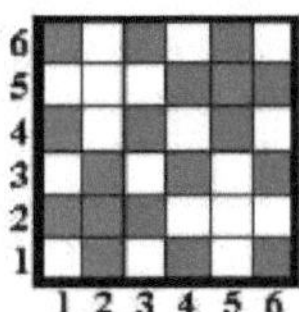

Figure 10. Location of intersections within a weave rapport.

Each warp and weft yarn has a different number of overlap changes The warp yarns - the first, third, fourth and sixth have six overlap changes each, and the second and fifth have two overlap changes each. The weft yarns - first, third, fourth and sixth have six overlap changes each, and the second and fifth have two overlap changes each. Hence, on the warp the average number of overlaps will be: $t_{ocp} = \Sigma P / Ro = 28/6 = 4.7$. For weft, the average number of overlaps will be: $t_{ycp} = \Sigma v / Ry = 28/6 = 4.7$. However, those weft yarns, which have a higher number of crossings, will be in the fabric in a more tense state and thus affect the length of the warp rapport. Therefore, when calculating the length of the warp and weft rapport.

it is necessary to take the maximum number of warp weft and weft warp crossings per yarn, i.e. t_{omax} and t_{ymax} . In this example, for the false openwork weave, the number of warp weft crossings t_{ymax} = 6 and warp weft crossings t_{omax} = 6. The length of the rapport is determined by the formulas: on warp $L_{Ro} = t_{ymax} l_o + (R_o - t_{ymax}) d_o$, on weft $L_{Ry} = t_{omax} l_y + (R_y - t_{omax}) d_y$, where d_o - diameter of warp thread in the fabric; l_o - distance between warp threads in the places of their intersections with weft; where d_y - diameter of weft thread in the fabric; l_y - distance between weft threads in the places of their intersections with warp. Substituting all values into the formulas for determining the maximum possible fabric density by warp and weft through the average distance between warp and weft yarns in the fabric, we obtain, $P'_o=100R_o / t_{ymax} l_o + (R_o - t_{ymax}) d_o$; $P'_y=100R_y / t_{omax} l_y + (R_y - t_{omax}) d_y$.

The filling factor of the fabric in warp and weft is determined according to the formulas:

based on $K_{Ho}=P_o[t_{ymax} l_o+(R_o - t_{ymax})d_o]/100R_o$, по утку $K_{Hy}=P_y[t_{omax} l_y+(R_y - t_{omax}) d_y]/100R_y$.

For plain weave fabric with $R_o=R_y$ = 2 and $t_{omax}=t_{ymax}$ = 2. Fibre filling coefficient is determined by the formulas: $K_{Ho}=P_o l_o /100$; $K_{Hy} =P_y l_y /100$.

Fibre fill factor of the fabric

$$K_H=K_{Ho} K_{Hy}. \quad (2.18).$$

This coefficient can be used to characterise the tension of the fabric production on the weaving machine. The more this coefficient approaches one, the tenser the fabric production on the machine. Consequently, the ratio of the actual density to the maximum density is characterised by the filling of the fabric with fibrous material, which takes into account the fabric density, the linear yarn density and the weave of the fabric. The latter, apart from the weave rapport, also takes into account the number of thread transitions of one system with respect to another system. Above we have considered the methodology of determining the number of thread transitions for falsely openwork weaves and on the basis of this index we have given the methodology of calculating the coefficient of filling with fibrous material and working out of warp and weft yarns by the average number of thread transitions. Yield is one of the main parameters, which allows to estimate the conditions of fabric production on the machine in the first approximation. The yield is used to determine the raw material consumption. Yield is influenced by raw material type, yarn linear density, yarn cross-sectional

shape, weave, warp and weft density, order of construction phase, threading and weaving parameters, *physical*, mechanical and rheological properties of the yarns. Therefore, the criterion for assessing the fabric structure can be taken as the yarn work-out. The difference between the lengths of the yarns before weaving and in the weave is referred to as workmanship.
Experimentally determine the yield on the basis of

$$ao=(Lo - LT)100/Lo \ \% \ (2.19).$$

Weft yield $ay=(Ly - B_T)100/Ly \ \%$ (2.20).

where: Lo, Ly - warp and weft lengths before weaving; L_T , B_T - length and width of the fabric.

Theoretically, the yarn yield in the fabric is determined by the mutual arrangement of the yarns in the fabric with some assumptions: the shape and cross-sectional area of the yarns along the entire length of the fabric are constant; the distance between the centres of the yarns of one system in the places where they intersect with the yarns of another system and in overlaps is proportional to the fill factor of the corresponding system.

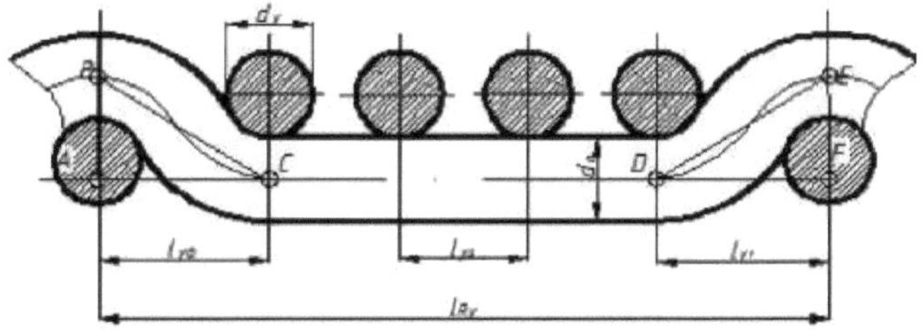

Fig.3.1 The arrangement of threads in the fabric.

The length of one warp thread in (Lo) is determined (Fig.10) by the length of the broken line BCDE, and the length of the fabric by the straight line ACDF. Workmanship on the basis of

$$a_o = (L_{BCDE} - L_{ACDF})/ L_{BCDE} - 100$$

$$ao = (BC+CD+DE-AC-CD-DF) / BC+CD+DE- 100 =$$

$$= BC+DE-AC\ DE / BC+CD+DE- 100$$

where: BC, DE - areas where the warp yarns cross the weft yarns

$$BC = DE = \sqrt{AC^2 + AB^2} = \sqrt{l_{y\phi}^2 + h_o^2}$$

l_{yf} - actual distance between the centres of weft yarns in the places where they intersect the main yarn, i.e. taking into account the order of the fabric structure phase and the actual weft density of the fabric

$$l_{y\phi} = l_y / K_{Hy} = (\sqrt{(d_o + d_y)^2 - hy^2}) / K_{Hy} \qquad (2.21)$$

The length of the straight part of the yarn is equal to the sum of the distances between the weft yarns

$CD = l_{y2}(R_y - t_{ocp}) = (R_y - t_{ocp})\, d_y / K_{Hy}$

where : l_{u2} - distance between weft yarns in places of long main overlap t_{ocp}, t_{ycp}

Substituting the obtained expressions, we determine the yarn work in the fabric by warp

$$a_o = [t_{ocp}\,(\sqrt{l_{y\phi}^2 + h_o^2} - l_{y\phi}) / t_{ocp}(\sqrt{l_{y\phi}^2 + h_o^2}) + (R_y - t_{ocp})\, d_y/K_{Hy}] \cdot 100 \quad (2.22)$$

Similarly determine the weft seam allowance in a fabric

$$a_o = [t_{ycp}\,(\sqrt{l_{o\phi}^2 + h_y^2} - l_{o\phi}) / t_{ycp}(\sqrt{l_{o\phi}^2 + h_y^2}) + (R_o - t_{ycp})\, d_o/K_{Ho}] \cdot 100 \quad (2.23)$$

where: K_{Ho} , K_{Hy} - coefficients of filling of fabric with fibrous material on the basis and on the weft.

The warp yield of the fabric is determined:

$$a_o' = \frac{t_o \cdot \sqrt{l_{y\phi}^2 + h_o^2} - l_{y\phi}}{t_o \cdot \sqrt{l_{y\phi}^2 + h_o^2} + (R_y - t_o)\frac{d_y}{K_{Hy}}} \cdot 100 \quad (2.24)$$

Weft yarn work in the fabric:

$$a_y' = \frac{t_y \cdot \sqrt{l_{o\phi}^2 + h_y^2} - l_{o\phi}}{t_y \cdot \sqrt{l_{o\phi}^2 + h_y^2} + (R_o - t_y)\frac{d_o}{K_{Ho}}} \cdot 100 \quad (2.25)$$

where: R_0, R_y - warp and weft rapport; t_o, t_y - number of weft crossings by warp and weft by weft; Tsf, - actual distance between the centres of warp yarns in the places where they are crossed by weft; l_{yf} - actual distance between the centres of weft yarns in the places of their intersection with warp; h_o, I_y - height of bending waves of warp and weft yarns; d_o, d_y , - diameters of warp and weft yarns; K_{Ho} , K_{Hy} - coefficients of fabric filling with fibrous material in warp and weft.

Fabric thickness determines such fabric properties as breathability, thermal insulation, drapeability (ability to form pleats), stiffness, elasticity and others. Different types of yarns with different fibre composition, degree of twist, fluffiness or smoothness determine greater or lesser fabric thickness. The type of weave determines the thickness of the fabric; the longer the overlap (planking) of the yarns and the larger the yarn rapport in the fabric, the greater the number of yarns placed under the planking, thickening the fabric. Also affecting the thickness of the fabric is the density of the fabric and the degree of bending of the threads in their weave, the greater their values, the thicker the fabric. Theoretically, the thickness of single-ply fabrics ranges from two to three warp and weft thread diameters.

The third chapter contains technological research of falsely openwork fabrics. The production of falsely openwork fabrics was carried out on the machine "Somet", the number of reams in the dressing -12 reams, the density of warp and weft 18 n/cm, the linear density of the main threads 10x2 tex, the linear density of weft threads varied from 10 tex to 50 tex, the dressing tension varied for one warp thread from 6 cN to 21 cN, for weft from 5 cN to 30 cN. The same variants were used to produce fabrics with the use of capron yarn in the weft. The fabric samples were produced in ten variants (Fig. 3.1) of falsely openwork weaves, taking into account the weave rapports in the warp and weft, the number of crossings of yarns of one system by another system, the fibre filling factor in the warp and weft. Analytical studies of the variants of falsely openwork weaves were carried out in order to study the influence of the type of weave, linear density of weft yarns of the weft type of falsely openwork fabrics and to compare them with experimental results.

Analytically, the average yarn yield was determined by our proposed formulae for warp and weft:

$$a_o = \frac{100}{R_o}\sum_{i=1}^{n}\frac{t_o(\sqrt{l_{y\phi}^2+h_o^2}-l_{y\phi}}{t_o\sqrt{l_{y\phi}^2+h_o^2}+(R_y-t_o)\frac{d_y}{K_{Hy}}} \quad (3.1) \qquad a_y = \frac{100}{R_y}\sum_{i=1}^{n}\frac{t_y(\sqrt{l_{o\phi}^2+h_y^2}-l_{o\phi}}{t_y\sqrt{l_{o\phi}^2+h_y^2}+(R_o-t_y)\frac{d_o}{K_{Ho}}} \quad (3.2)$$

where: t_o, t_y - the number of warp and weft intersections of equally intertwining yarns within the fabric rapport for each weave motif; l_{of}, l_{yf} - the actual distance between the centres of yarns (warp - weft) at their intersections for each weave motif; h_o, h_y - height of the bending wave of the warp and weft yarns, respectively; R_o, R_y - weave ratio of the warp and weft yarns, respectively; n - number of equally intertwining yarns of each weave motif, the sum of these numbers is equal to the weave ratio of the fabric;

$(R_y - t_o)\frac{d_y}{K_{Hy}}$ - length of the rectilinear part of the thread within;

$(R_o - t_y)\frac{d_o}{K_{Ho}}$ - The weft and warp threads, respectively.

Ten variants of falsely openwork weaves have been developed and are shown in Fig. 3.1. Each variant in the weave motif has its own within the weave report on the warp and weft, the number of intersections of warp weft and weft warp threads, the number of equally intertwining threads of each weave motif, the sum of these numbers is equal to the weave ratio of the fabric. Table 3.1 according to Fig. 3.1. presents the warp and weft rapports, the number of intersections, the number of equally intertwining threads, the yarn processing for ten variants of weaves, the calculation is carried out according to the formulas (3.1) and (3.2). We also suggest that for convenience in the calculation, the crossings should be indicated by numerical indices, which after multiplication by the number two will show the number of thread transitions for each thread.

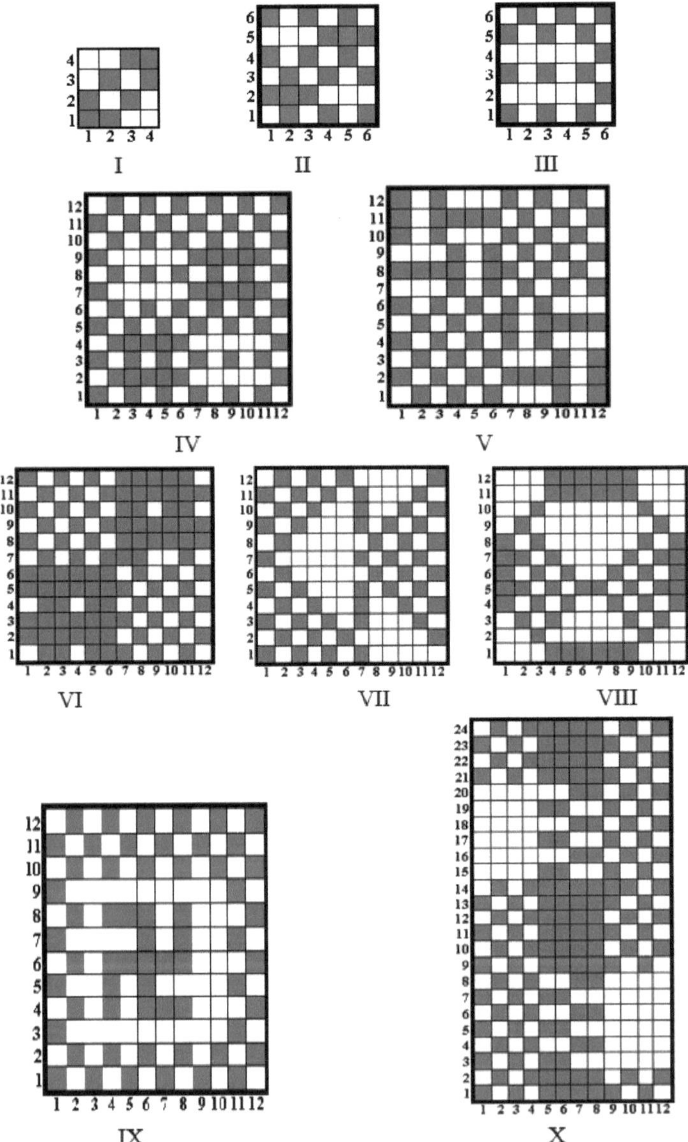

Fig.3.1 Variants of falsely openwork weaves.

The calculations carried out at linear density of weft T_y = 30 tex of cotton yarn and capron yarn according to the formulas (3.1) and (3.2) show that at the unchanged (the same) fabric rapport and the *same number of* crossings (variants II and III of weave) the average working of warp and weft yarns are the *same*. For variants IV and V (Fig. 3.1) at $R_o = R_y$ = 12: in the first case $t_{o2} = t_{y2}$ = 4 at $\boldsymbol{n}$ = 4, and $t_{o6} = t_{y6}$ = 12 at $\boldsymbol{n}$ = 6, the average weft and warp ***yarn*** processing a_o = 4,2%, a_y = 4,1%; in the second case $t_{o4} = t_{y4}$ = 8 at $\boldsymbol{n}$ = 4 and $t_{o5} = t_{y5}$ = 10 at $\boldsymbol{n}$ = 8 the average processing a_o = 3,8%, a_y = 3,5%. As we can see the number of intersections in the rapport is the same, but in the first case. These intersections have extreme values $t_{o2} = t_{y2}$ and $t_{o6} = t_{y6}$, and in the second case the average values $t_{o4} = t_{y4}$ and $t_{o5} = t_{y5}$ in the interval from t_1 ***to*** t_6, which leads to a decrease in the values of workability on the base by 11% and on the weft by 14%. It also follows that increasing the number of intersections of variants VI and IX within the rapport leads to an increase in the workmanship of warp and weft yarns by 23%. Consequently, the proposed calculation of the yarn yield takes into account the scattering of the values of the intersections of yarns at the same rapport and the same average intersections. The comparison of calculations of formulas (3.1) and (3.2) was carried out with formulas (3.3) and (3.4), besides, the influence of the type of cotton and capron yarns, linear density of weft yarns on the filling factor of fabrics and the yarn working of falsely openwork weaves for weft warp yarns was investigated:

$$a_o = \frac{t_{ocp}\left(\sqrt{l_{y\phi}^2 + h_o^2} - l_{y\phi}\right)}{t_{ocp}\sqrt{l_{y\phi}^2 + h_o^2} + (R_y - t_o)\frac{d_y}{K_{Hy}}} \quad (3.3) \qquad a_y = \frac{t_{ycp}\left(\sqrt{l_{o\phi}^2 + h_y^2} - l_{o\phi}\right)}{t_{ycp}\sqrt{l_{o\phi}^2 + h_y^2} + (R_o - t_y)\frac{d_o}{K_{Ho}}} \quad (3.4)$$

Table 3.1.

Results of calculation by formulas (3.1) and (3.2). parameters of tissue structure.

Options\ indicators

варианты / показатели	I	II	III	IV	V	VI	VII	VIII	IX	X
R_o	4	6	6	12	12	12	12	12	12	12
R_y	4	6	6	12	12	12	12	12	12	24
t_{o1}/n_1	2/2	2/2	2/2					2/2		
t_{o2}/n_2	4/2			4/4			4/2	4/2	4/2	
t_{o3}/n_3		6/4	6/4			6/7	6/2	6/2	6/1	
t_{o4}/n_4					8/4	8/3	8/3	8/4	8/3	

t_{o5}/n_5					$10/8$	$10/2$	$10/2$	$10/2$	$10/2$	
t_{o6}/n_6				$12/8$			$12/3$		$12/4$	$12/4$
t_{y1}/n_1	$2/2$	$2/2$	$2/2$					$2/3$		
t_{y2}/n_2	$4/2$	$6/4$	$6/4$	$4/4$				$4/3$	$4/2$	
t_{y3}/n_3						$6/7$		$6/2$	$6/1$	$6/12$
t_{y4}/n_4					$8/4$	$8/3$	$8/6$	$8/2$	$8/4$	$8/12$
t_{y5}/n_5					$10/8$	$10/6$	$10/6$	$10/2$		
t_{y6}/n_6				$12/8$					$12/5$	
t_{o9}/n_9										$18/8$
a_{o1}	3,2	2,3	2,3					1,1		
a_{o2}	5,1			2,3			2,2	2,0	2,3	
a_{o3}		5,1	5,1			2,7	3,0	2,8	3,2	
a_{o4}					3,4	3,4	3,7	3,4	3,9	
a_{o5}					3,9	3,9	4,3	4,0	4,5	
a_{o6}				5,1			4,8		5,0	3,0
a_{o9}										3,9
a_{ocp}	4,2	4,2	4,2	4,2	3,8	3,1	3,7	2,8	4,0	3,6
a_{y1}	3,2	2,5	2,5							
a_{y2}	4,8			2,5				1,6	2,4	
a_{y3}		5,0	5,0			2,7		2,2	3,2	2,6
a_{y4}					3,2	3,2	3,8	2,6	3,8	3,1
a_{y5}					3,7	3,7	4,3	3,0		
a_{y6}				4,9					4,7	
a_{ycp}	4,0	4,2	4,2	4,1	3,5	3,0	4,1	1,9	3,9	2,7

Fill factor in warp and weft:

$$K_{Ho} = \frac{P_o(d_oR_o+d_yt_{ycp})}{10R_o} \quad (3.5) \qquad K_{Hy} = \frac{P_y(d_yR_y+d_ot_{ocp})}{10R_y} \quad (3.6)$$

Table 3.2 shows the values of the average number of crossings on the warp $\mathbf{t_{ocp}}$ and on the weft $\mathbf{t_{ycp}}$, filling and working coefficients of yarns on the warp and on the weft for ten variants of weaves made of cotton and capron weft yarns with linear density $T_y = 30$ tex.

Table 3.2.

Results of calculation by formulae (3.3) and (3.4). parameters of tissue structure.

Options	*Ro*	*Ry*	*tocp*	*tycp*	Cotton weft				Capron duck			
					Кно	*KHy*	*ao*	*ay*	*Кно*	*KHy*	*ao*	*ay*
I	4	4	3,0	3,0	0,558	0,576	4,3	3,9	0,588	0,612	5,0	4,4
II	6	6	4,7	4,7	0,569	0,581	4,4	4,5	0,601	0,622	5,0	3,8
III	6	6	4,7	4,7	0,569	0,581	4,4	4,5	0,601	0,622	5,0	3,8
IV	12	12	9,3	9,3	0,565	0,578	4,4	4,3	0,598	0,62	5,2	4,6

V	12	12	9,3	9,3	0,565	0,578	4,4	4,3	0,598	0,62	5,2	4,6
VI	12	12	7,2	7,2	0,504	0,527	3,1	3,0	0,529	0,568	3,9	3,0
VII	12	12	8,3	9,0	0,556	0,554	3,8	4,1	0,588	0,595	4,3	4,6
VIII	12	12	6,3	5,5	0,454	0,535	2,9	2,0	0,473	0,546	3,6	1,8
IX	12	12	8,8	8,8	0,551	0,567	4,1	4,0	0,581	0,608	4,9	4,2
X	12	24	16	7	0,499	0,547	3,6	2,9	0,522	0,588	4,8	4,0

The calculations carried out according to formulae (3.3) and (3.4) show that for the same rapport and intersections (weave variants II, III, IV and V), the average warp and weft yarn yields are the same for both cotton and capron weft fabrics. Although, according to Table 3.1, the values of variants IV and V are different, due to the different variation of the intersection values **t**.

Table 3.3 summarises the effect of linear density of the weft (capron and cotton) on the fill factor and on the yarn retention.

Table 3.3.

Influence of weft linear density on fill factor and on yarn work-up.

№	Linear density of weft yarn tex	Fill factor		Average yarn throughput	
		On the basis of K_{Ho}	On duck ***KHy***	On the basis of ***ao***	On the duck ***ow***
1	20	0,523/0,544	0,523/0,550	3,7/4,2	3,7/3,9
2	30	0,569/0,588	0,581/0,612	4,4/5,0	4,5/4,4
3	40	0,611/0,612	0,635/0,683	5,9/7,4	4,4/4,3
4	50	0,647/0,691	0,682/0,737	7,5/8,5	4,3/4,4
5	60	0,681/0,729	0,725/0,786	9,2/10,5	4,0/4,6

With the increase in the linear density of weft yarns from 20 to 60 tex, the tension of fabric production increases, the working out of warp yarns increases, and the working out of (cotton) weft yarns decreases, and the working out of (capron) weft yarns remains almost unchanged. Experimental research confirms the analytical research in that the character of the change in the workmanship in all variants is ideintic, i.e. the workmanship is influenced by the number of crossings in the fabric rapport and the type of raw materials used in the weft. The discrepancies in absolute values of workmanship are caused by the fact that the formulas (3.1 and 3.2) do not take into account technological modes of fabric production (warp and weft tension, size of overstep, size of shed, etc.). Table 3.4 shows the results of the study of processing in the fabric samples obtained experimentally. The workmanship of warp and weft yarns was determined on the worked fabric samples according to the following methodology. A thread was taken out of a 10x10 cm fabric sample and the straightened thread was measured and then calculated according to the formulas (2.19) and (2.20). Such measurements and calculations are made at least five times. Then, the numerical characteristics of the workmanship and their errors were determined.

1 .Average value of the yarn processing: $\overline{Y}=\frac{1}{m}\sum_{i=1}^{m} Y_i$

2 .Dispersion: $S^2\{Y\}=\frac{1}{m-1}\sum_{i=1}(Y_i-\overline{Y})^2$

3 . Mean square deviation: $S\{Y\}=\sqrt{S^2\{Y\}}$

4 .coefficient of variation $C\{Y\}=\frac{S\{Y\}}{\overline{Y}}100$

5 .Absolute confidence error of the mean: $E\{\overline{Y}\}=S\{\overline{Y}\}\frac{t_T}{\sqrt{m}}$

where: *t_T {PD = 0.95, /= m-1=5-1-1=4} = 2.776, the* quantile of Student's distribution.

6 .Relative confidence error of the mean value $\delta\{\overline{Y}\}=C\{Y\}\frac{t_T}{\sqrt{m}}$ where: *t_T {PD = 0.95, /= m-1=5-1-1=4} = 2.776,* Student's distribution quantile. The error of the obtained values did not exceed 5%.

Table 3.4.

Results of a study of workmanship in fabric samples.

Sample weave variations	Yarn yield %			
	Cotton weft		Capron duck	
	On the basis of	On the duck	On the basis of	On the duck
I	6,2±0,1	5,7±0,2	6,7±0,3	2,8±0,1
II	6,0±0,3	6,0±0,3	6,6±0,2	4,0±0,2
III	6,5±0,3	6,4±0,3	6,7±0,3	3,3±0,2
IV	6,5±0,2	6,1 ±0,2	6,6±0,3	3,4±0,1
V	6,7±0,1	5,8±0,3	6,5±0,3	3,4±0,1
VI	4,5±0,1	4,3 ±0,1	5,8±0,3	2,8±0,1
VII	6,2±0,2	6,3±0,2	6,0±0,3	3,0±0,1
VIII	4,7±0,2	4,3±0,3	5,4±0,3	2,3 ±0,1
IX	5,8±0,2	5,7±0,2	6,1 ±0,3	3,0±0,1
X	4,8±0,3	4,7±0,2	6,0±0,3	3,1±0,1

Comparing the results of Tables 3.2 and 3.4, it can be noted that the character of yarn yield change in all variants is ideal, i.e. the yield is influenced by the number of crossings in the fabric rapport and the type of raw materials used in the weft. We have large differences in the absolute values of yarns yields. This is due to the fact that the yield calculated by the formulas (3.3), (3.4), (3.5) and (3.6) does not take into account the technological modes of fabric production (warp and weft tension, the amount of overstep, the size of the shed, etc.).

Tables 3.5 and 3.6 show the results of the investigation of the effect of warp and weft filling tension on yarn yield.

Table 3.5.

Influence of filling tension of warp yarns on workmanship.

№	Filling tension of the warp yarns cN/thread	Filling tension of weft yarns cN/thread	Yarn yield %	
			basis	ducks
1	6	10	8,5 / 8,7	4,8 / 2,7
2	10	10	7,3 / 7,4	5,0 / 3,2
3	12	10	6,6 / 7,1	5,5 / 3,5
4	16	10	6,0 / 6,6	6,0 / 4,0
5	19	10	5,4 / 6,1	6,3 / 4,5
6	21	10	5,0 / 5,4	6,8 / 4,9

- numerator for cotton weft - denominator for capron weft.

It is established that with changing the filling tension of warp threads from 6 cN to 21 cN (for a single warp thread), other conditions being equal: the warp yield decreases for fabrics made of cotton wefts by 1.7 times, and for fabrics made of capron wefts by 1.6 times; the weft yield increases respectively for cotton wefts by 1.4 times, and for capron wefts by 1.8 times. Changing the filling tension of weft yarns from 5 cN to 30 cN yarn processing: on the warp increases when producing fabric from cotton weft by 1.5 times, from kapron weft by 1.7 times; on the weft decreases respectively in cotton weft by 1.6 times and in kapron weft by 1.9 times.

Table 3.6.

Influence of filling tension of weft yarns on workmanship.

№	Filling tension of weft yarns cN/thread	Filling tension of warp yarns cN/thread	Yarn yield %	
			basis	ducks
1	5	16	5,1 / 5,3	6,6 / 4,9
2	10	16	6,0 / 6,6	6,0 / 4,0
3	15	16	6,5 / 6,9	5,4 / 3,5
4	20	16	6,9 / 7,7	4,9 / 3,1
5	25	16	7,3 / 8,6	4,4 / 2,8
6	30	16	7,9 / 9,1	4,0 / 2,5

- Numerator for cotton weft.

- the denominator for the kapron point.

The technical calculation of the falsely openwork fabric is also given. The second variant of false openwork weave (Fig. 3.1) was accepted as the base fabric for the technical calculation. Initial data for the calculation: T_o = 10 x 2 tex; T_u = 30 tex; $R_o = R_u$ = 18 n/cm; $N_б$ = 60 teeth/dm; $z_φ$ = 3 threads per reed tooth; a_o = 4.4 % ; a_y = 4.5 % ; K_{no} = 0.569; K_{nu} = 0,581; d_{cT} = 250mm ; $d_φ$ = 800mm ; H = 1900mm; n = 440 min^{-1} ; P_{rem} = 12; B_c = 175cm; $B_{зб}$ = 183,3cm; $n_{K\ rom}$ = 32; $n_K\ p$ = 4; p_o = 3300. The weaving machine from Somet Super Excel is adopted:

Weight of warp yarns.

$$M_o = \frac{n_o * T_o * 100}{(1-\frac{a_o}{100}) * 10^6} = \frac{3304 * 10x2 * 100}{(1-\frac{4,4}{100}) * 10^6} = 6,91кг$$

Weight of weft yarns:

$$M_y = \frac{10 * P_y * l_y * T_y * 100}{10^6} = \frac{10 * 180 * 1,853 * 30 * 100}{10^6} = 10,01кг$$

Weight per linear metre of fabric:

$$M_{п.м.} = \frac{M_o + M_y}{100} * 1000 = \frac{6,91 + 10,01}{100} * 1000 = 169гр/м;$$

Surface density of the tissue:

$$M_м^2 = \frac{M_{п.м.}}{B_c} = \frac{169}{1,75} = 97гр/м^2$$

Weight of cut leno yarns in 100m of warp fabric:

$$M_{onep} = \frac{n_{кром} * T_o * 100}{(1-\frac{a_{nepoe}}{100}) * 10^6} = \frac{32 * 20 * 100}{(1-\frac{40}{100}) * 10^6} = 0,11кг$$

Weight of cut leno yarns in 100m of weft fabric:

$$M_{ynep} = \frac{10 * P_y * l_y * T_y * 100}{10^6} = \frac{10 * 180 * 0,1 * 30 * 100}{10^6} = 0,54кг$$

The length of the warp coming on one piece of fabric:

$$l_{ok} = \frac{100 l_K}{100 - a_o} = \frac{100 * 50}{100 - 4,4} = 52,3м$$

Associated weaving bulk weight

$$G_H = \frac{L_H * n_o * T_o}{10^6} = \frac{5234 * 3300 * 20}{10^6} = 346кг$$

Associated lengths on the spool

$L_б = L_H\ n_б + 120 = 5234 * 20 = 104800$ м.

Associated bobbin weight

$$G_б = \frac{L_б * T_o}{1000} = \frac{104800 * 20}{1000} = 2,1кг$$

Belt weaving losses

$$У_c = \frac{(l_1 * a + l_2 * K) * 1{,}5}{L_{TK}} = \frac{(2 * 164{,}8 + 1 * 34{,}5) * 1{,}5}{17272200} * 100 = 0{,}003\%$$

$$L_{TK} = L_H * n_o = 5234 * 3300 = 17272200м.$$

$$a = \frac{L_H * H_o}{L_б} = 164{,}8$$

$$K = O_б * L_H * n_o * 10^{-6} = 2 * 5234 * 3300 * 10^{-6} = 34{,}5$$

The hijinks of going through the basics

$$У_{np} = \frac{l_n}{L_H} * 100 = \frac{1}{5234} * 100 = 0{,}02\%$$

Ugars of binding basics

$$У_{уз} = \frac{l_1 + l_2}{L_H} * 100 = \frac{1+1}{5234} * 100 = 0{,}04\%$$

Total percentage of punching and tying losses

$$У_n = \frac{0{,}02 * 10 + 0{,}04 * 90}{100} = 0{,}037\%$$

Yields on the warp loom

$$У_o = \left(\frac{l_1 + l_2}{L_H - L_n} + O_o\right) * 100 = \left(\frac{1{,}5 + 1{,}5}{5234 - 1} + 0{,}016\right) * 100 = 1{,}66\%$$

$$O_o = \frac{M_{onep}}{M_o} = \frac{0{,}11}{6{,}91} = 0{,}016$$

Losses on the weft loom

$$У_y = \left(\frac{l_1 + l_2 * K + l_3}{L_б} + O_y\right) * 100 = \left(\frac{2 + 1 * 1{,}6 + 10{,}5}{104800} + 0{,}054\right) * 100 = 5{,}40\%$$

$$K = \frac{O_б * У_б}{Py * ly} = \frac{0{,}05 * 104800}{1800 * 1{,}853} = 1{,}6$$

$$O_y = \frac{0{,}54}{10{,}01} = 0{,}054$$

Total percentage of carbon monoxide by base

$$У_{oc} = У_c + У_n + У_o = 0{,}003 + 0{,}037 + 1{,}66 = 1{,}7\%$$

Warp consumption per 100m of fabric, including shrinkage

$$M_o^1 = M_o * \left(1 + \frac{У_{oc}}{100}\right) = 6{,}91 * \left(1 + \frac{1{,}7}{100}\right) = 7{,}03кг$$

Weft consumption per 100m of fabric including fading

$$M_y^1 = M_y * \left(1 + \frac{У_y}{100}\right) = 10{,}01 * \left(1 + \frac{5{,}4}{100}\right) = 10{,}55кг$$

Technological process of false openwork fabric production

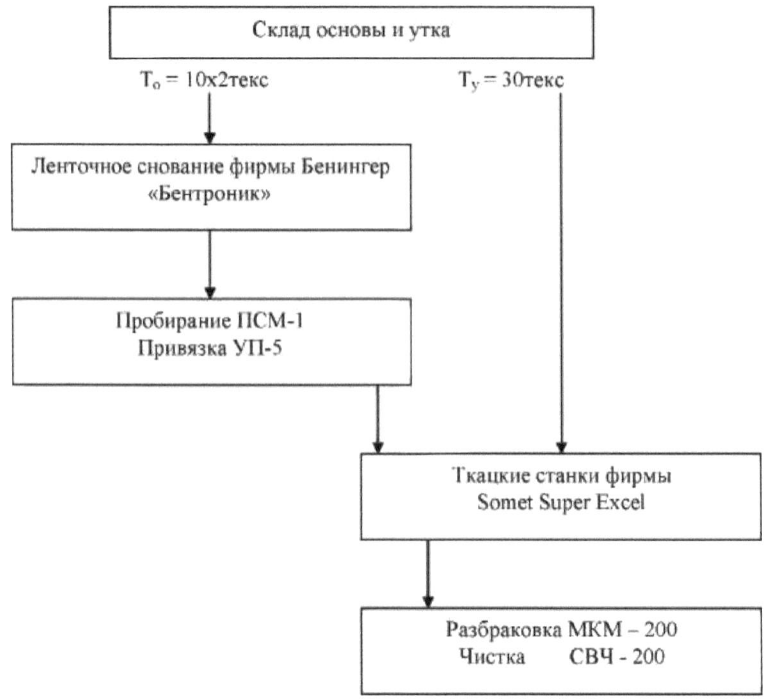

Appropriate technological parameters for the transitions are selected.

In the fourth chapter physical-mechanical, hygienic and consumer properties of falsely openwork fabrics are given and the influence of the number of thread transitions within the fabric rapport and the type of raw materials used in the weft on these properties is shown. Various textile materials are widely used in the production of footwear as linings, cushions and shoe tops. They must have a complex of physical-mechanical and hygienic properties and wear resistance. During operation, they are subjected to repeated mechanical effects (stretching, rescue, bending, etc.). As the top of shoes often use synthetic materials that worsen the hygienic properties of shoes, so the use of natural materials, in particular fabrics, significantly improve these properties, taking into account the climatic conditions of the region. The developed samples of fabrics (10 variants) were tested for physical-mechanical properties (stretching, elongation, abrasion) and hygienic properties (air permeability) in the certification centre at TITLP on modern devices according to the developed methodology of laboratory research of fabrics.

The results of these studies are presented in Table 4.1, where the numerator shows the values of fabric production indices with capron weft. Table 4.1 shows that with increasing number of warp (t_{ocp}) and weft (t_{ycp}) transitions: breaking load on warp and weft increases; fabric abrasion increases; fabric air permeability decreases, the exception is variant II, as the weave forms a net in the fabric. Capron weft increases breaking load, breaking elongation and breathability of the fabric. It is recommended to use falsely openwork fabrics of the II variant as a shoe upper, as this fabric has good indicators of physical-mechanical, hygienic and consumer properties.

Table 4.1.

Physicomechanical and hygienic properties of falsely openwork fabrics.

№	Binding options	Ro	Ry	tocp	ty p^c	On the basis of		On the duck		erasure	Air permeability
						Rroh	Iro	P Rroo	Iru		
1	I	4	4	3,0	3,0	323	618	618	13,5	69	54
						328	1015	1015	16,6	38	127
2	II	6	6	4,7	4,7	74	12,7	671	12,9	74	162
						"3	9,7	1671	32,6	28	260
3	III	6	6	4,7	4,7	363	12,7	654	12,5	64	66
						328	9,7	1634	32,6	22	127
4	IV	12	12	9,3	9,3	371	12,5	630	12,5	61	88
						305	8,7	1655	31,6	18	156
5	V	12	12	9,3	9,3	371	12,5	626	13,9	57	89
						3й	12,7	1634	31,8	19	169
6	VI	12	12	7,2	7,2	275	10,4	11,9	11,9	35	113
						274	12,0	33,7	33,7	16	237
7	VII	12	12	8,3	9,0	617	10,4	617	12,6	47	123
						1614	8,7	1614	27,8	15	203
8	VIII	12	12	6,3	5,5	236	10,7	560	10,2	21	134
						269	14,4	1551	15,9	12	253
9	IX	12	12	8,8	8,8	320	10,1	599	12,2	49	127
						308	10,1	1604	22,2	17	221
10	X	12	24	16	7,0	384	13,0	574	11,8	53	97
						361	15,2	1570	31,2	16	182

The analysis of Table 4.1 shows that physical-mechanical and hygienic properties of the fabric are influenced by the number of warp and weft yarns transitions within the rapport, as well as by the type of raw materials used in the weft (numerator - cotton, denominator - kapron). Tensile strength data, 38 abrasion and air permeability clearly illustrate that with increasing number of yarn transitions (t_{ocp} and t_{ycp}): breaking load on warp and weft increases; fabric abrasion increases; fabric air permeability decreases, with the

exception of variant II, as the weave forms a net type of fabric.
The use of capron weft leads to an increase in breaking load, elongation and breathability, and to a decrease in abrasion compared to fabrics in which cotton weft is used, due to the physical and mechanical properties of capron thread (breaking load, low coefficient of friction, elongation, etc.). It is expedient to use false-journal fabrics of the II variant as the upper of footwear, as this fabric has good indicators of physical and mechanical properties, hygienic properties and consumer properties.

Optimisation of the weaving process for the production of falsely openwork **fabric is carried out**

In the textile industry the following optimization problems are solved by linear programming methods: optimisation of fibre blend composition by its cost; optimisation of blend composition by fibre irregularity (their length, linear density, breaking load, etc.); determination of the optimal assortment of yarn, fabric and knitted fabrics produced by enterprises on the available equipment; optimal distribution of production of a given assortment of yarn, fabric and knitted fabrics on the available equipment; selection of the optimal way of transporting raw materials (p
Practically no solution to an optimisation problem can be expressed by a finite algebraic expression, it can be obtained only as a result of successive algebraic and logical operations. The method of solving such problems is described by some algorithms. An algorithm is an unambiguously defined sequence of actions leading to the solution of a problem (in a particular case, an algorithm allows to establish that the problem has no solution or has no final solution). An algorithm refers to computational methods of solution, so it is always based on numerical procedures. A number of requirements are imposed on algorithms, the fulfilment of which characterises its capabilities: completeness, meaning the unambiguity of the rules prescribed by the algorithm, the presence of a practical way of constructing the initial trial solution, if necessary, the need to anticipate all situations that may arise in the process of calculations for different initial data; the area of applicability is determined by the class of problems to be solved, the simplicity of determining whether the problem belongs to this class, the computational procedures of the computational procedures, the time-consuming nature of the computational procedures. During the study, directly controls the course of optimisation on the model, agrees it with his ideas about the process, makes corrections in the course of optimisation (changes the search area, parameters or structure of the model, introduces new or excludes existing

constraints, modifies the target function, etc.), significantly accelerating and increasing the efficiency of the solution. Thus, when solving an optimisation problem, it is possible to combine the advantages of the computer and mathematical model (speed of solution, systematic approach) with the experience and intuition of the decision maker. The first solution is a mathematical concept: an optimal solution turns the target function to extremum, which serves as an indicator of the effectiveness of the decision. The efficient solution is an engineering concept: an efficient solution is understood as the best solution among all possible solutions, but its definition is not limited to formal methods of optimisation, but requires taking into account the experience and intuition of the researchers making the decision. Therefore, optimal and efficient solutions most often do not coincide. This difference plays a fundamental role at the stage of implementation of the results of the solution of optimisation problems. If the difference is large, the value of the optimal solution found with the help of a mathematical optimisation model and the effort spent on optimisation are close to zero. The degree of difference is determined mainly by the adequacy of the used model to the optimised process. Building a complete process model and thus ensuring adequacy is very difficult. However,
usually the values of the controlled variables close to the optimum are known before optimisation, so the gain from solving the optimisation problem is small compared to the effort and time spent. The efficiency of optimisation with the help of models can be increased if the adequacy of the model to the optimised process is achieved not by complicating it, i.e. by ensuring its completeness, but by adapting the model to the changing conditions of the process that are not reflected in the original model. The interests of a researcher making a decision are usually not reduced to an optimal or efficient solution. It is more important to know the consequences of deviating from the efficient solution, in particular because of the difference between the optimal and efficient solutions. Questions of this kind relate to analysing for the sensitivity of the process to variations of parameters under near-optimal conditions. The use of a model makes it much easier to solve this problem.
Three main parameters, i.e. three main independent variables, were chosen in this work: x_1 - warp filling tension, cN; x_2 - size of the scarp, mm; x_3 - position of the scalp relative to the sternum in height, mm. It is established that warp thread breakage at low values of the filling tension increases due to the increase of the surge. Then, as the warp tension increases, the breakage rate decreases and increases again as the warp tension increases further due to

warp overstressing. The size of the backstop affects the conditions of the weft fringe and, as a consequence, the size of the fringe. As the backstop increases, the weft yarn run-in conditions improve and the size of the surf strip decreases. Therefore, when producing fabrics with a higher filling ratio, a larger backstop is recommended. As the backstop decreases, the size of the bump increases and a dynamic shock phenomenon occurs, which can increase the breakage of warp yarns. The degree of movement of the weft on the warp yarns at the moment of surf depends on the tension of the main yarns in the upper and lower parts of the shed. When the tension of the shed branches is different, more favourable conditions are created for the weft yarn surfing, i.e. the conditions of fabric formation are improved and the warp yarn breakage is reduced. On the other hand, large deviations from the symmetrical shed level can create a weakening of tension in one branch and increase it in the other, i.e. lead to breakage of weakly tensioned warp yarns. The difference in tension between the upper and lower branches of the shed can be controlled by the height of the cliff. The selected factors meet all the requirements of the theory of mathematical planning of experiment: there is no interchangeability of factors, they can be measured by available means, they can be varied within a sufficiently wide range of minimum and maximum values and taken with the necessary accuracy. As for the other technological parameters of the machine tool fuelling, all of them were constant during the experiment. The selection of intervals and values of factors for five levels of variation was carried out from the consideration of technological capabilities of machine tool dressing, which is presented in Table 1.

Table 4.2...

Levels of variation of factors

Factors	Levels of variation					Interval
	-1,682	-1,0	0	+1,0	+1,682	
x_1-filling warp tension, cN	13	16	20	24	27	4
X2-value of offset, mm	7	10	15	20	23	5
x_3 - rock position relative to the sternum, mm	-15	-10	0	+10	+25	15

Rotatable Central Composition Experiment (RCCE)

we carry out to describe the stationary section of the response surface and the experiment carried out on the selected matrix allows us to obtain a second-order mathematical model describing the influence of factors x_1, x_2 , x_3 on the selected optimisation parameters of the following form

$$y = в_0 + в_1x_1 + в_2x_2 + в_3x_3 + в_{12}x_1x_2 + в_{23}x_2x_3 + в_{13}x_1x_3 + в_{11}x_1^2 + в_{22}x_2^2 + в_{33}x_3^2 \quad (4.1)$$

where: c_0 , B_i , $c^\wedge$, c_a - regression coefficient; c_0 - free term; c_1, c_2 , c - regression coefficients for linear terms; c_{12} , c_{23} , c_{13} - coefficients for factor interaction; c_{11}, c_{22}, c_{33} - regression coefficients of squared terms.

Table *4.3* shows the RCCE planning matrix. Calculation of regression coefficients and subsequent calculations were carried out in the following sequence.

$$в_0 = g_1 \sum_{u=1}^{N} \bar{y}_u - g_2 \sum_{i=1}^{M} \sum_{u=1}^{N} x_{iu}^2 \bar{y}_u$$

$$в_0 = 0{,}1663(3{,}44) - 0{,}0568(8{,}076) = 0{,}113$$

Table 4.3.

RCCE planning matrix

Order of randomisation	Experience number	Factors			Optimisation criterion Yu warp breaks	$(Y_U-Y_R)^2$
		X1	X2	X3		
20	1	+	+	+	0,28	0,000049
19	2	+	+	-	0,20	0,000061
18	3	+	-	+	0,27	0,00001
17	4	+	-	-	0,33	0,000026
16	5	-	+	+	0,36	0,000052
15	6	-	+	-	0,43	0,000073
14	7	-	-	+	0,29	0,000057
13	8	-	-	-	0,49	0,000016
1	9	0	0	0	0,26	0,0001
2	10	0	0	0	0,25	0,0001
3	11	0	0	0	0,27	0,0001
4	12	0	0	0	0,26	0
5	13	0	0	0	0,25	0,0001
6	14	0	0	0	0,25	0,0001
11	15	+1,682	0	0	0,50	0,000010
12	16	-1,682	0	0	0,63	0,000100
9	17	0	+1,682	0	0,32	0,000292
10	18	0	-1,682	0	0,40	0,000015
7	19	0	0	+1,682	0,18	0,000090
8	20	0	0	-1,682	0,20	0,000034

$$в_1 = g_2 \sum_{u=1}^{N} x_{iu} \overline{Y}_u$$

$$в_1 = 0{,}0732 \cdot (-0{,}455) = -0{,}033$$

$$в_2 = g_3 \sum_{u=1}^{N} x_2 \overline{Y}_u$$

$$в_2 = 0{,}0732 \cdot (-0{,}11) = -0{,}008$$

$$в_3 = g_3 \sum_{u=1}^{N} x_3 \overline{Y}_u$$

$$в_3 = 0{,}0732 \cdot (-0{,}15) = -0{,}011$$

The pairwise interaction coefficients of the regression equation were determined using the formula:

$$в_{ij} = g_u \sum_{u=1}^{N} x_{iu} \cdot x_{ju} \cdot \overline{Y}_u$$

$$в_{12} = 0{,}125 \cdot (-0{,}12) = -0{,}015$$
$$в_{13} = 0{,}125 \cdot (0{,}18) = 0{,}023$$
$$в_{23} = 0{,}125 \cdot (0{,}2) = 0{,}025$$

The coefficients of the quadratic terms of the regression equation are determined by the formula:

$$в_{ij} = g_6 \sum_{u=1}^{N} x_u^2 \cdot \overline{Y}_u + g_6 \sum_{i=1}^{M} \sum_{u=1}^{N} x_{iu}^2 \cdot \overline{Y}_u - g_2 \sum_{u=1}^{N} Y_u$$

$$в_{11} = 0{,}0625 \cdot (3{,}58) + 0{,}0069 \cdot (8{,}076) - 0{,}0568 \cdot (3{,}44) = 0{,}084$$
$$в_{22} = 0{,}0625 \cdot (2{,}53) + 0{,}0069 \cdot (8{,}076) - 0{,}0568 \cdot (3{,}44) = 0{,}018$$
$$в_{33} = 0{,}0625 \cdot (1{,}965) + 0{,}0069 \cdot (8{,}076) - 0{,}0569 \cdot (3{,}44) = -0{,}017$$

According to the method of processing the experimental results, the variance of the output parameter in the experiment or the variance of reproducibility is determined:

$$S^2\{\overline{y}\} = S_u^2\{y\} = \frac{1}{N_u - 1} \sum_{U_u=1}^{N_u=6} \left(y_{u_u} - \overline{y}_u\right)^2$$

$$S^2\{\overline{y}\} = \frac{1}{5} \cdot 0{,}000134 = 0{,}00003 \qquad \overline{y}_u = 0{,}113$$

Next, the variance of the regression coefficients is determined:

$$S^2\{в_0\} = g_1 \cdot S^2\{\overline{y}\}$$

$$S^2\{в_0^0\} = 0{,}1663 \cdot 0{,}00003 = 0{,}000005 \quad S\{в_0^0\} = 0{,}0022$$

$$S^2\{в_i\} = g_3 \cdot S^2\{\overline{y}\}$$

$$S^2\{в_i^0\} = 0{,}0732 \cdot 0{,}00003 = 0{,}0000022 \qquad S\{в_i^0\} = 0{,}0015$$

$$S^2\{в_{ij}\} = g_4 \cdot S^2\{\overline{y}\}$$

$$S^2\{в_{ij}^0\} = 0{,}125 \cdot 0{,}00003 = 0{,}0000038 \quad S\{в_{ij}^0\} = 0{,}0019$$

$$S^2\{в_{ii}\} = g_7 \cdot S^2\{\overline{y}\}$$

$$S^2\{в_{ii}^0\} = 0{,}0695 \cdot 0{,}00003 = 0{,}0000021 \qquad S\{в_{ii}^0\} = 0{,}0014$$

$$t_R = \frac{|b|}{S\{b\}}$$

$$t_R\{b_0\} = \frac{0{,}113}{0{,}0022} = 51{,}36$$

$$t_R = \frac{|b_i|}{S\{b_i\}}$$

$$t_R\{b_1\} = \frac{0{,}033}{0{,}0015} = 22$$

$$t_R\{b_2\} = \frac{0{,}008}{0{,}0015} = 5{,}33$$

$$t_R\{b_3\} = \frac{0{,}011}{0{,}0015} = 7{,}33$$

$$t_R\{b_{ij}\} = \frac{|b_{ij}|}{S\{b_{ij}\}}$$

$$t_R\{b_{12}\} = \frac{0{,}015}{0{,}0019} = 7{,}9$$

$$t_R\{b_{13}\} = \frac{0{,}023}{0{,}0019} = 12{,}1$$

$$t_R\{b_{23}\} = \frac{0{,}025}{0{,}0019} = 13{,}2$$

To determine the significance of the regression coefficients, Student's criteria are used

$$t_R\{b_{ii}\} = \frac{|b_{ii}|}{S\{b_{ii}\}}$$

$$t_R\{b_{11}\} = \frac{0{,}084}{0{,}0014} = 60{,}0$$

$$t_R\{b_{22}\} = \frac{0{,}018}{0{,}0014} = 12{,}9$$

$$t_R\{b_{33}\} = \frac{0{,}017}{0{,}0014} = 12{,}15$$

Tabular value Student's criterion

$$t_T\left[P_D = 0{,}95; \quad fS_y^2 \; 6-1=5\right] = 2{,}571$$

Regression coefficients are significant if tR>tT. In this case, all the coefficients are significant. The mathematical model describing the dependence of breakage on the selected factors is as follows

$$y_R = 0{,}112 - 0{,}032X_1 - 0{,}008X_2 - 0{,}011X_3 - 0{,}015X_1 \cdot X_2 + 0{,}024X_1 \cdot X_3 + 0{,}023X_2 \cdot X_3 + 0{,}082X_1^2 + 0{,}017X_2^2 - 0{,}016X_3^2 \quad (4.2)$$

To test the hypothesis of adequacy of the obtained model, we use Fisher's criterion, the calculated value of which is compared with the tabulated F_T . If $F_R < F_T$, then with probability P_D the hypothesis of adequacy of the obtained model is not rejected.

The calculated values of Fisher's criterion are equal:

$$F_R = \frac{S_{a\partial}}{S_E}$$

$$S_E = \sum_{u=1}^{N_u} (y_u - \bar{y})^2$$

$$S_R = \sum_{u=1}^{N} (\bar{y} - y_R)^2$$

$$S_{a\partial} = S_R - S_E$$

$$S_E = 0{,}00008$$

$$S_R = 0{,}00019$$

$$S_{a\partial} = 0{,}00019 - 0{,}00008 = 0{,}00011$$

$$F_R = \frac{0{,}00011}{0{,}00008} = 1{,}38$$

Tabular value of Fisher's criterion

$$F_T[P_D = 0{,}95; \quad f_1 = N(m-1) = 20, \quad f_2 = N - N_k = 14] = 2{,}28$$

Since $F_R < F_T$, the hypothesis of adequacy of the obtained models is not rejected at confidence level $P_D = 0.95$.

In order to determine the optimal parameters of fabric production, we will evaluate the efficiency of technological experiment by means of cuts: y= f (x) at constant x_2 , x_3 ; y= f (x_2) at constant x_1 , x_3 ; y= f (x_3) at constant x_1 , x_2 . Tables 4.4 to 4.6 summarise the results of the breakage calculations from the input factors. As can be seen all equations are parabola equation. The analysis of curves (Figs. 4.1-4.2) plotted by the obtained equations shows that the change of y from x_1 and x_2 has the form of concave parabolas. (Figures 4.1, 4.2, 4.4). The influence of x3 (position of the scalp relative to the breast of the loom Fig. 4.3) on y is represented by a convex parabola with minimum y values at x_3 equal to - 1.68 and + 1.68 respectively. Consequently, during the production of this fabric it is advisable to raise the scala as much as possible, or to lower it as much as possible in relation to the sternum, and raising the scala (x_3= +1.68), as shown in Fig. 4.4 leads to the lowest breakage. Separately plotted curve of change of breakage y from the position of the offset x2 at zero value x1 (tension of the main threads) and the maximum lifted rock x_3= +1,68 shows (see Fig. 4.4) that at x2=0 it is possible to reduce the breakage of threads in 2 times, ie the parameters will have the following values: the tension of the main threads - 20 cN (per 1 thread); the value of the offset - 15 mm.; the position of the rock relative to the sternum - (+ 25) mm. At these values of parameters breakage of main threads will not exceed 0,05 breaks per 1m.

Table 4.4.

Calculation results y=f(x_1) at constant x2 and x_3

№	Constant values	Yarn breakage *at the* value of factor x1 variables

	of factors	-1,682	-1	0	+1	+1,682
1	x2=-1, xz=-1	0,465	0,283	0,158	0,201	0,327
2	x2=-1, Xz=0	0,407	0,241	0,139	0,205	0,347
3	x2=-1, Xz=1	0,261	0,164	0,085	0,174	0.zz1
4	x2=0, Xz=-1	0,385	0,247	0,107	0,135	0,251
5	x2=0, Xz=0	0,407	0,23	0,113	0,164	0,295
6	x2=0, Xz=1	0,340	0,179	0,085	0,159	0,306
7	x2=1, xz=-1	0,449	0,247	0,092	0,105	0,211
8	x2=1, xz=0	0,391	0,225	0,123	0,189	0.zz1
9	x2=1, Xz=1	0,400	0,229	0,12	0,179	0,316

Table 4.5.

Calculation results $y=f(x_2)$ at constant x_1 and x_3

№	Constant values of factors	Warp thread breakage *y* variables x value of factor x_2				
		-1,68	-1	0	+ 1	+1,68
1	x1=-1, xz=-1	0,328	0,283	0,247	0,247	0,268
2	x1=-1, xz=0	0,269	0,241	0,230	0,255	0,293
3	x1=-1, xz=1	0,176	0,165	0,179	0,229	0,284
4	x1=0, xz=-1	0,214	0,158	0,107	0,092	0,102
5	x1=0, xz=0	0,177	0,139	0,113	0,123	0,151
6	x1=0, xz=1	0,107	0,086	0,085	0,12	0,165
7	x1=1, xz=-1	0.z1z	0,247	0,181	0,151	0,151
8	x1=1, xz=0	0,254	0,205	0,164	0,159	0,176
9	x1=1, xz=1	0,207	0,175	0,159	0,179	0,213
10	x1=0,xz=1.68	0,241	0,115	0,046	0,098	0,155

Table 4.6.

Calculation results $y=f(x_3)$ at constant x1 and x2

№	Constant values of factors	Warp thread breakage *y* variables x factor $value_3$				
		-1,68	-1	0	+ 1	+1,68
1	x1=-1, X2=-1	0,190	0,283	0,241	0,165	0,094
2	x1=-1, X2=0	0,239	0,247	0,23	0,179	0,125
3	x1=-1, X2=1	0,222	0,247	0,255	0,229	0,192
4	x1=0, X2=-1	0,151	0,158	0,139	0,086	0,051
5	x1=0, X2=0	0,115	0,107	0,103	0,085	0,077
6	x1=0, X2=1	0,052	0,092	0,123	0,120	0,098
7	x1=1, X2=-1	0,179	0,201	0,205	0,175	0,135
8	x1=1, X2=0	0,096	0,135	0,164	0,159	0,136
9	x1=1, X2=1	0,06	0,100	0,159	0,179	0,173

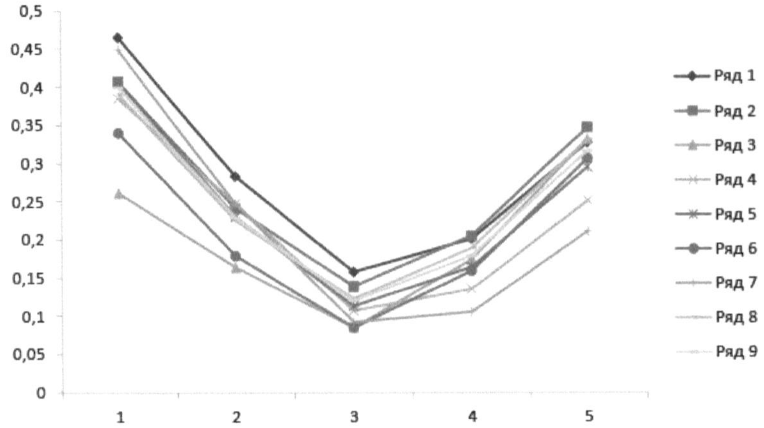

Row1 - X2 = -1 ; xz = -1 Row2- X2 = -1 ; xz = 0 Row3 - X2 = -1 ; xz = 1
Row4 - $x_2 = 0$; x_3 = -1 Row5- $x_2 = 0$; $x_3 = 0$ Row6 - $x_2 = 0$; x_3 = 1
Row7 - x2 = 1 ; xz = -1 Row8- x2 = 1 ; xz = 0 Row9 - x2 = 1 ; xz = 1

Figure 4.1. Influence of warp filling tension on thread breakage

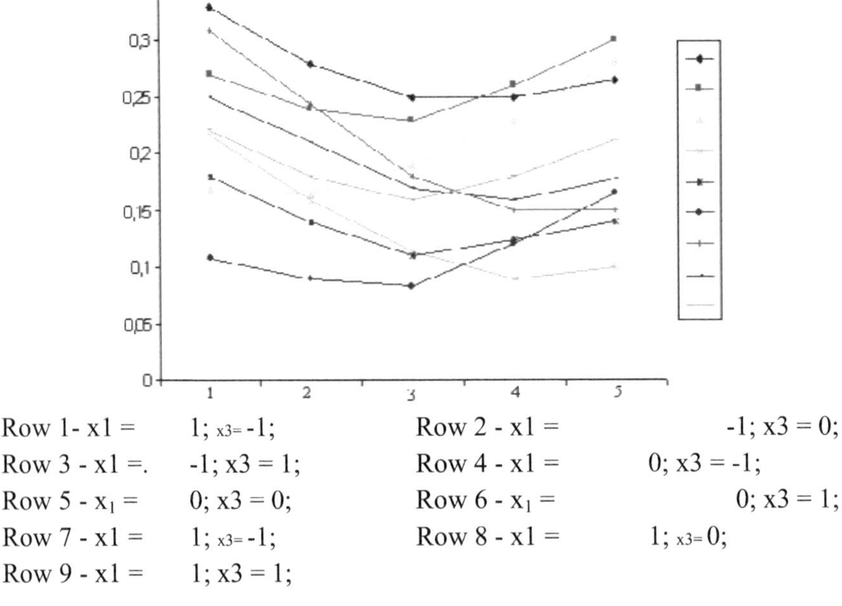

Row 1- x1 = 1; x3= -1; Row 2 - x1 = -1; x3 = 0;
Row 3 - x1 =. -1; x3 = 1; Row 4 - x1 = 0; x3 = -1;
Row 5 - x_1 = 0; x3 = 0; Row 6 - x_1 = 0; x3 = 1;
Row 7 - x1 = 1; x3= -1; Row 8 - x1 = 1; x3= 0;
Row 9 - x1 = 1; x3 = 1;

Fig.4.2 Influence of the size of the gap on thread breakage

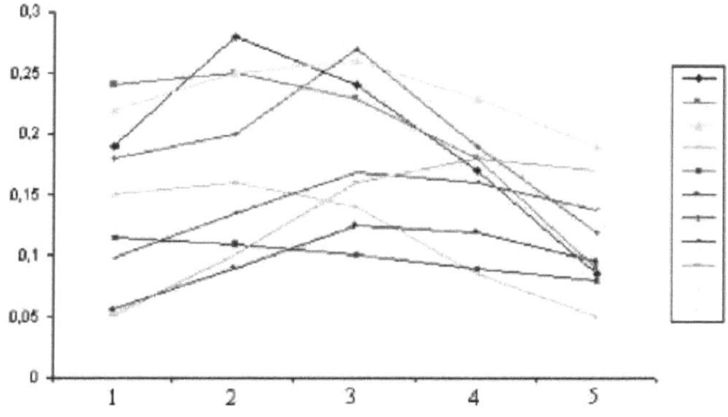

Row 1- x1 =-1 ; x2 = -1; Row 2 - x1 = -1; x2 =0 ;
Row 3 - x1 =-1 ; x2 = 1; Row 4 - x1 =0 ; x2=-1 ;
Row 5 - x1 =0 ; x2 = 0; Row 6 - x1 =0 ; x2 =1 ;
Row 7 - x1 =1 ; x2= -1; Row 8 - x1 =1 ; x2=0 ;
Row 9 - x1 =1 ; x2 = 1;

Fig.4.3 Effect of scalo position on thread breakage

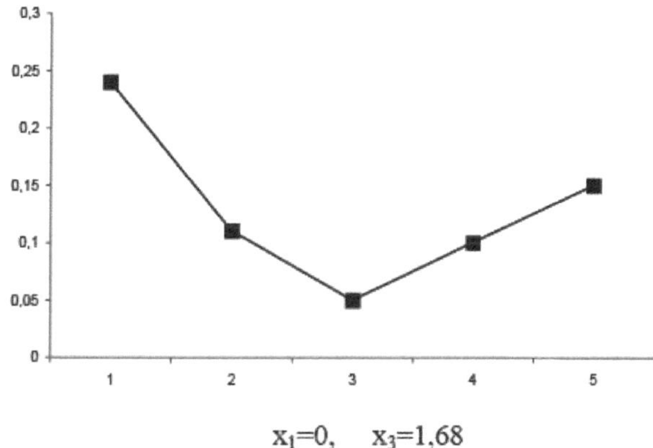

$x_1=0, \quad x_3=1,68$

Fig. 4.4 Influence of the amount of overstep on thread breakage

Determination of the RCCE includes the following stages: conducting a preliminary experiment; planning the PFE; finding the conditions for conducting the PFE; conducting the main experiment according to the experiment planning matrices; processing the experimental results; analysing the obtained model. The preliminary experiment is conducted to determine the accuracy of measurement of the output parameter and the confidence level of its measurements and to test the hypothesis of normality of distribution of random variables. PFE planning includes: compilation of the planning matrix, planning of repeated experiments and randomisation of experiments. In factor planning, unlike traditional (single-factor) planning of the experiment, the value of regression coefficients can be used to judge the influence of not only each input factor, but also their interaction, i.e. the change in the influence of one factor when the second factor is transferred to another level. Having the experiment matrix and experimental conditions, the researcher conducts the main experiment by entering the obtained repeated samples in the experiment table of the values of the output parameter. Since the weaving process is non-stationary in time and a large number of experiments distort the results due to different disturbances in the process, randomisation of experiments was applied in planning the experiment. In this work, the central composite method of planning the experiment of the second order was adopted, which gives the possibility of detailed study, description and optimisation of the weaving process in the investigated area of optimisation. The selection of intervals and factor values for the five levels of variation was made from the consideration of the technological capabilities of loom fuelling, which is presented in Table 4.2. Processing of PFE results, 49 includes the following operations. Exclusion of sharply distinguished values. Testing the hypothesis of homogeneity of dispersions in the matrix experiments using the Kochren criterion. As a result of calculations, a regression multifactor model is obtained, which, however, is not the final model of the studied process. Having obtained the mathematical model (4.2), we analyse it. The values of regression coefficients characterise the contribution of the corresponding factor to the value of the output parameter when the factor moves from the basic level to the upper or lower one. After checking the significance of regression coefficients, this RMFM specifies the test of significance of regression coefficients. For this purpose, Student's criterion is used, the calculated value of which is compared with the tabular one. If the calculated value is greater than the tabular value, the hypothesis of significance of regression coefficients is not rejected. In case of insignificance of any regression coefficient it can be discarded without recalculation of all other coefficients. In our case, all in the model of equation (4.2) are significant. To roughly assess the adequacy of a linear mathematical model, the following indirect signs are used: if at least one interaction regression coefficient is significant, the linear model cannot be adequate; if the hypothesis of adequacy is rejected, it is necessary to proceed to the description of the process by a second-order polynomial on the basis of another type of experiment or, if possible, to conduct an experiment with a smaller interval of variation of factor levels. However, the reduction of the variation interval leads to the need to increase the number of repeated experiments and to a decrease in the values of regression coefficients. The adequacy of the obtained model was checked using Fisher's criterion. The obtained model (4.2) is adequate. The greater the coefficient in the regression equation (4.2), the higher the effect of this factor, i.e. the stronger the influence of the factor on the output parameter. Thus, by the value of the regression coefficients

in the model we can rank the factors by the strength of their influence on the output parameter. The sign in front of the regression coefficient determines the nature of the influence of the factor on the output parameter. Factors whose coefficients have a plus sign increase the value of the output parameter, and those with a minus sign reduce it. On the basis of mathematical model the estimation of technological experiment was carried out by means of slices: $Y = f(x_1)$ at constant x_2 , x_3 ; $Y = f(x_2)$ at constant x_1 , x_3 ; $Y = f(x_3)$ at constant x_1 , x_2 . Analyses of graphs 1.- 3 show all equations are parabola equation, in Figures 4.1-4.2 the variation of **Y** from x1 and x2 has the appearance of concave parabolas. The effect of x3 (position of the scalp relative to the breast of the loom Fig.3) on **Y** is represented by a convex parabola with minimum y values at x_3 equal to - 1.68 and + 1.68 respectively. Consequently, during the production of this fabric it is advisable to raise the scala as much as possible, or to lower it as much as possible in relation to the sternum, and raising the scala ($x3 = +1.68$), as shown in Fig. 4.1 leads to the lowest breakage. Separately plotted curve of change of breakage y from the position of the offset x_2 at zero value x1 (tension of the main threads) and the maximum lifted rock $x_3 = +1{,}68$ shows (see Fig. 4.1) that at $x_2 = 0$ it is possible to reduce the breakage of threads in 2 times, i.e. the parameters will have the following values: tension of the main threads - 20 cN (per 1 thread); the value of the offset - 15 mm.; the position of the rock relative to the sternum - (+ 25) mm. At these values of parameters breakage of main threads should not exceed 0,05 breaks per 1m.

In addition, the design of shoe fabrics for shoe insoles has been carried out. In shoe manufacturing, various textile materials are widely used as linings, cushions (insoles) in some cases and shoe tops, the latter being investigated in the previous chapters. The desired insole thickness (2.5-5.0 mm) can be obtained in three ways.

In the first method, several layers of fabric are glued together. However, in this case the air permeability and hygroscopicity of the fabric are reduced and the process of shoe manufacturing becomes more difficult.

In the second method, several layers of fabric are stitched or woven together. In this case, labour costs increase and the technology of producing insoles becomes more complicated.

In the third method, the thickness of the insoles is regulated by the use of yarns of high linear density in the fabric. This causes good hygroscopicity and breathability of insoles, simplifies the technology of manufacturing insoles, the surface of the insoles have reliefs, which are in contact with the soles of the feet massage the feet when walking.

According to Su Jok therapeutic methodology, the feet are an exact reflection of the entire human body, so massaging painful points (zones, areas) can eliminate pain and provide therapeutic (preventive) effects on the sick body and the body as a whole.

At the present stage medical workers and scientists treat many ailments - massaging with seeds, acupuncture, magnets, moxa (burning with wormwood sticks), etc. We offer an insole with therapeutic properties differing in that on the surface of the fabric there is a relief in the form of seeds, and depending on the contact zones in the relief can be used fibres of silk, flax, yucca, cotton, kenaf and etc. In the work at the first stage the task of designing fabrics according to the given thickness on the basis of computer technology was set. To solve the task, an algorithm and a programme for calculating the thickness of the fabric were compiled. In this case the following initial parameters were set - fabric thickness ***Ttk***, yarns buckling coefficient in the vertical direction of warp η_{OB} and weft η_{YB} , yarns buckling coefficient in the horizontal direction of warp $\eta_{OГ}$ and weft $\eta_{YГ}$, weave pattern of fabric on warp Ro and on weft R_y, coefficients determining the warp Co and weft Su diameters, the number of warp warp weft $\mathbf{t_0}$ and weft warp $\mathbf{t_y}$, the coefficient of the ratio of thread diameters *Kd, the* coefficient of the ratio of fabric densities ***Kp***, presented in Table 4.7.

Table 4.7.

Defined parameters of the structure of the designed fabric.

Ttk	L ov	7 uv	L og	7 ug.	R o	Ry,	So	Soo	t.	t_y	Kd	Kp
4,5	0,8	0,7	1,2	1,3	2	2	1,25	1,25	2	2	1	2,1

Then we determined the limits and interval of possible values of the average yarn diameter d_{cp} and the average yarn density in the fabric P_{cp}. We calculated the height of bending wave of weft yarns ***hu*** and warp yarns ho, geometric density of warp *lo,* or weft ***1****y* , the sum of vertical diameters of one yarn system and horizontal diameters of another yarn system ***A***, the maximum possible height of bending waves of weft h_{ymax} or warp ho_{max} , the calculated fabric thickness ***Ttkr***. The results of calculations are shown in Table 2.Comparison of the values of the specified fabric thickness Ttk with the calculated values Ttkr shows the closest variant is the second one, i.e. $T_{ТК} = T_{ТКр}$ 4,5 ~ 4,43. Further we have determined the diameters of warp threads ***do*** and ***dy of*** weft, linear density of warp threads T_o and weft threads T_y , fabric density by warp P_0 and by weft P_y , which are presented in Table 4.8. The analysis of Table 4.8 and Table 4.9 shows that geometric density decreases hyperbolically with increasing average fabric density, and also with increasing average fabric density and average yarn diameter the height of yarn bending wave, technological density and fabric thickness increase linearly. For the production on the hand loom, the designed fabric of the second variant ($T_o = T_y$ =50x48 tex, P_o =42 threads/dm, P_y =20 threads/dm) and additional variants, which are presented in Table 4, are taken as a base.10. Fibre composition of the yarn in all variants, cotton twisted yarn in warp and weft, plain weave, reed number N_6 = 20 teeth/dm, number of yarns per reed tooth z =2 and z =2.5.

Table 4.8.

Results of fabric thickness calculation

№	dcp	*R* cf cf	*h* ***y***	*lo*	A	*h* ***o***	TtKO	TtKp
1.	1,875	0,615	0,907	1,200	3,750	2,662	4,162	4,162
2.	1,988	0,581	0,962	1,270	3,975	2,840	4,430	4,430
3.	2,100	0,549	1,016	1,344	4,200	3,016	4,696	4,696
4.	2,213	0,522	1,071	1,414	4,425	3,192	4,962	4,962
5.	2,325	0,496	1,125	1,488	4,650	3,367	5,227	5,227
6.	2,438	0,473	1,179	1,561	4,875	3,542	5,492	5,492
7.	2,550	0,452	1,234	1,633	5,100	3,716	5,756	5,756
8.	2,663	0,433	1,288	1,705	5,325	3,889	6,019	6,019
9.	2,775	0,416	1,343	1,774	5,550	4,062	6,282	6,282
10.	2,888	0,400	1,397	1,845	5,775	4,235	6,545	6,545
11.	3,000	0,385	1,452	1,1917	6,000	4,408	6,808	6,808

Table 4.9.

Calculation results of fabric technological density

№	*dy* mm	***do*** mm	*T* o o tex	*T* *1* ***at*** tex	*P* ***o*** thread/mm	*P* yarn/mm
1.	1,875	1,875	2250	2250	0,416	0,198

2.	1,988	1,988	2528	2528	0,394	0,188
3.	2,100	2,100	2822	2822	0,374	0,177
4.	2,213	2,213	3133	3133	0,354	0,169
5.	2,325	2,325	3460	3460	0,336	0,160
6.	2,438	2,438	3802	3802	0,320	0,153
7.	2,550	2,550	4162	4162	0,307	0,146
8.	2,663	2,663	4537	4537	0,293	0,140
9.	2,775	2,775	4928	4928	0,282	0,134
10.	2,888	2,888	5336	5336	0,271	0,129
11.	3,000	3,000	5760	5760	0,261	0,124

Table 4.10.

Technical characterisation of shoe fabric samples

No. Fabric samples	Linear yarn density, tex		Thread density per 10cm of fabric		Surface density of fabric gr/m^2
	basis	ducks	On the basis of	On the duck	
1.	50x48	50x24	43	23	1350
2.	50x48	50x48	42	20	1530
3.	50x48	50x48	26	31	1390
4.	50x48	50x24	27	34	1080

The results of Table 4.11 show that each of the produced samples has different quality indicators. Therefore, for the convenience of their comparison, the rank complex evaluation of processing results was used, which are shown in Table 4.12.

Table 4.11.

Physical, mechanical and hygienic properties of footwear fabrics

No. Fabric samples	The breaking load of the strands , H		Water resistance mm. water column	Air - Capacity cm /cm^{32} . sec	Wear resistance cycles	gesture bone mn/cm^2	Wettability %
	basis	ducks					
1.	450	430	10	7,50	5000	419899	22,9
2.	480	520	12	2,63	5000	458967	20,0
3.	420	480	11	3,27	5000	445687	21,9
4.	430	410	8	22,12	5000	385625	25,1

Table 4.12.

Ranking assessment of shoe fabric properties

No. Fabric samples	Breaking load of yarns, N		Water resistance, mm. water column, mm. water column .	Air resistance, cm^3 / cm^2 sec	Wear resistance, cycles	Stiffness, mn/cm^2	Wet-ability, %	Sum of ranking scores
	Os nova	ducks						
1.	2	3	3	2	1	3	2	16
2.	1	1	1	4	1	1	4	13
3.	4	2	2	3	1	2	3	17
4.	3	4	4	1	1	4	1	18

According to the ranking results, the second sample, which has the lowest sum of ranks, is recognised as the best. This sample is recommended as cushioning (insoles) for footwear.

CONCLUSION

A leno means of reducing the tension of stockinette threads at the moment of transition of leno eyes relative to stockinette threads is developed. The method of calculation of the working out for each thread within the fabric rapport is offered. The variants of samples of falsely openwork fabrics are developed and worked out, in which it is shown that the increase in the number of thread crossings within the rapport leads to the increase in the working of warp and weft threads in the fabric. Tension of production of falsely openwork fabrics on the machines depends on the type of weft yarn. At processing of capron weft this index increases by 10-12% in relation to cotton weft. The workmanship on the warp increases and on the weft decreases when the linear density of weft thread increases from 20 tex to 60 tex and the filling tension increases from 5 cN. to 30 cN. per thread when laying capron weft. The warp yield decreases and weft yield increases when the filling tension of warp yarns is changed from 6 cN to 21 cN of a single yarn. The character of the change in yarn yield obtained analytically and experimentally is identical. The discrepancies of absolute values of working time are caused by technological modes, not taken into account in the analytical calculations. The technical calculation of the fabric is made and the rational technological process of preparation of false openwork fabrics production is chosen. With the increase in the number of thread transitions within the rapport, the breaking load on the base and weft, fabric abrasion increases, and the air permeability of the fabric decreases. Capron weft in fabric increases breaking load on weft, breaking elongation on weft and air permeability of fabric, but reduces abrasion of fabric. It is recommended to use falsely openwork fabric of II variant, which has good physical-mechanical, hygienic and consumer indicators of properties of falsely openwork fabrics, for shoe tops. The mathematical model of main threads breakage depending on their parameters, in particular on the tension of main threads, the size of the scalp and the position of the scalp relative to the sternum has been developed. The selection of intervals and values of factors for five levels of variation was carried out, taking into account the technological possibilities of machine filling. Estimation of research results was carried out with the help of Smirnov-Grabs, Kochren, Fisher, Student's criteria. On the basis of the received mathematical model the estimation of technological experiment was carried out with the help of slices On weaving machines at production of falsely openwork fabrics it is recommended to establish optimal parameters: tension of main threads - 20 cN (on 1 thread); value of backstep - 15 mm.; position of the scala relative to the breast - (+25) mm. At these values of parameters breakage of the main threads will not exceed 0.05 breaks per 1m. In addition, such a combination of technological parameters allows you to better preserve the strength properties of the thread in the fabric. A new structure of insoles with therapeutic properties is developed, parameters of shoe fabrics structure are selected. An algorithm and a programme for calculating shoe fabrics by a given thickness have been developed. With the increase of the average density of the fabric the geometrical density hyperbolically decreases, and the technological density linearly increases. With increasing average yarn diameter, the bending wave height and fabric thickness increase linearly. Experimental samples of footwear fabrics are developed and their physical-mechanical and hygienic properties are investigated.

Based on the ranking scores, the second sample shoe fabric was recommended as an insole.

LITERATURE

1. HANDBOOK OF YARN PRODUCTION Technology, science and economics P R Lord, NCSU, USA 504 pages 244 x 172mm hardback July 2003.

2. WELLINGTON SEARS HANDBOOK OF INDUSTRIAL TEXTILES Edited by S Adanur Ars Textrina A Journal of Textiles and Costume, Winnipeg, Canada TS 1300 A77.

3. Patra, A.K. and Pattanayak, A.K. (2015) New varieties of denim fabrics. In: Paul, R., Ed., Denim fabric, Woodhead Publishing Limited, Cambridge, 483-506.

4. Elmogahzy, YE (2020) Performance characteristics of traditional textiles: denim and sportswear products. In: Elmogahzy, YE, Ed., Engineering Textiles, Elsevier, Amsterdam, 319-346.

5. Islam S., Parveen F., Urmi Z., Ahmed S., Islam S. Investigation of solutions to environmental pollution and workers' health problems caused by textile manufacturing operations. International Journal of Textile Research. 2020; 2: 1-21.

6. Islam S., Parvin F., Urmi Z., Ahmed S., Arifuzzaman M., Yasmin J., et al. Investigation of human health benefits, human comfort properties and environmental effects of natural sustainable textile fibres. European Journal of Physiotherapy and Rehabilitation Research. 2020; 1: 1-24.

7. Kadirova D.N., Daminov A.D.,Rahimhodjaev S.S., Technology of production of technical belts and the study of their properties. Scoups,2019/ 549-552 8.Martynova A.A., Vlasova N.A., Slostina G.L. Textbook for students of universities/. - M.: Izd. MSTU,1999/ 343str.

9 . Novikov N.G. About fabric structure and its design by means of geometrical method// Textile industry. 1946,№2,4,5,6,11.C.42.

10 .A.D. Daminov. On topology of interposition of warp and weft yarns in weaving overlaps No.3.2003.

11 A.D. Daminov. On the dependence of weave coefficient on the order of the fabric structure phase No.4.2004 p.26.

12 Nikolaev S.D. Forecasting of technological parameters of fabrics manufacturing of a given structure and development of their calculation methods. D. in Technical Sciences - M., MTI, 1989.

13 Khamraeva S.A. Increase of wear resistance of fabrics by optimisation of parameters of their formation. Author's abstract of dissertation....-doctor of technical sciences.- Tashkent, TITLP, 2010.

14 . Vlasov P. V. "Normalisation of the weaving process" Moscow, L. I., 1982.
15 Rakhimkhodjaev S.S. et al. Analytical research of yarn processing of shoe fabrics of false-journal weave. No.1.2007 p. 54.
16 Rakhimkhodjaev S.S. et al. Influence of some parameters on the structure of fabrics of false-journal weaves. No.2.2007 pp. 34.
17 . Kadirova D.N., Daminov A.D.,Rahimhodjaev S.S., Technology of production of technical belts and the study of their properties. Scoups,2019/ 549-552.
18 Tsybik-Dorzhieva A.. V. Estimation of technologicity of threads at manufacturing of fabrics of various weaves. Avtoref... kand sciences .Moscow 2009.
19 .Murad MM, Elshakankeri MH, Almetwalli A. A. Physical and tensile properties of cotton fabrics with different spandex content. J Am Sci 2012; 8: 567-72.
20 .Gersak J, Sain D, Wiley B. Investigation of relaxation phenomena in fabrics containing elastane yarns. Int J Cloth Sci Technol 2005; 17:188-99.
21 .Inozemtseva N. A.Development of the method of designing fabrics on the given order of the structure phase. Avtoref. kand sciences .Moscow - 2010.
22 . Kadirova D.N. Daminov A.D., Rakhimkhodjaev S.S. Technology, design and parameters of technical fabrics. Monograph, 2020. LAPLAMBERT ACADEMIC PUBLISHING, Mauritius. p-170.
23 Rakhimkhodjaev S.S., Kadyrova D .N. Theoretical bases of the process of Tissue formation. Textbook. Tashkent. TITLP. 2018.
24 . Bukaev P.T. Cotton weaving. M., Legprombytizdat, 1987.
25 R.Meredith.Physical methods of research of textile materials,M,Gizlegprom,1963.
26 A.G.Sevostyanov "Methods and means of research of technological processes". M. Light Industry. 1980 г.
27 . Raximxodjaev S.S , D.N.Qodirova To'qima loyialashning zamonaviy usullari. Darslik.-T.: Adabiyot uchqunlari. 2018-144b.
28 .Ortikov O.A., Rasulov H.Y., Kadirova D.N., Rakhimkhodjaev S.S. Optimisation of yarn tension on weaving machines with microspacers // Monograph, 2017. LAPLAMBERT ACADEMIC PUBLISHING, Mauritius. p-224.
29 . HANDBOOK OF WEAVING Edited by S Adanur, Department of

Textile Engineering, Auburn University, USA 440 pages 543 figures 68 tables 254 x 176mm hardback 2000.
30 Rasulov, S.S.Rakhimkhodjaev. Analytical research of a tension of the warp yarns for the cycle of work of the weaving loom. International Journal of Advanced Research in Science, Engineering and technology. Vol.5, Issue 10, October 2018. - pp. 7001-7005.
31 Aoki M. Introduction to Optimisation Methods. M., 1997.
32 .Nogin V. D., Protodyakonov I. O., Evlampiev I. I. Fundamentals of Optimisation Theory. M., 1986.
33 .Moiseev N. N. N., Ivanilov Y. P., Stolyarova E. M. Optimisation Methods. M. Methods of optimisation. M., 1978.
34 Rekleitis G., Reivindran A., Ragsdell K. Optimisation in engineering. M., 1986 (in 2 books).
35 .Ventzel E. S. Operations Research: Problems, Principles, Methodology. M., 1988.
36 .GOST 19196-73. Fabrics cotton shoe fabrics. Part 1 M., I.S. 1978.
37 .Park Jae-woo et al. Su jok treatment on the hand and foot. Minsk. Interfair, 1993.
38 .Rakhimkhodjaev S.S., Kadyrova D.N. Modern methods of fabric design. TITLP, Tashkent, 2006.

Printed by Books on Demand GmbH, Norderstedt / Germany